master fibonacci

MASTER FIBONACCI

THE MAN WHO CHANGED MATH

SHELLEY ALLEN, M.A.ED.

FIBONACCI INC.
fibonacci.com
2018

Published in the United States by
Fibonacci Inc.
1309 N Ave E
Shiner, Texas 77984

ISBN 978-1-5323-9349-5

Printed in the United States of America

CONTENTS

FOREWORD

My first recollection of the Fibonacci sequence (0,1,1,2,3,5,8,13...) and the quotient of its successive numbers, phi (0.618), also known as the Fibonacci constant or the Golden Ratio, occurred in high school math class. Like pi (π) or Euler's exponential constant, *e*, phi (φ) seemed meaningful only as a requirement for exams.

That phi persisted as an important mathematical concept through my engineering course of study in college was not particularly surprising. That it became a prevailing theme in a Humanities seminar on the artwork of Leonardo da Vinci, Michelangelo and the topic of what constitutes beauty, was so notable that I remember the class today, more than twenty years later. The notion that the Golden Ratio proposes an answer to the question of whether there is an objective measure of beauty has always been fascinating to me.

Later on, in various media, I casually noted references to the Fibonacci sequence in biology, geography, and other disciplines, but not until I took an earnest interest in commodities trading did I fully appreciate Fibonacci's ubiquity. In technical trading (a topic for another book) Fibonacci ratios serve as a useful predictor of market movements.

In 2018, I purchased fibonacci.com to serve as a resource for those, like me, interested in the applications of this remarkable sequence and ratio, and I asked Shelley Allen to pen an introductory book on the origins and prevalence of the Fibonacci sequence in art, music, nature, and other disciplines. In this well-researched arrangement, with numerous references and over 100 citations, Allen offers the reader a concise overview of the subject along with breadcrumbs for continued learning. Along the way, the reader will learn how Roman numerals were replaced in Europe, and how the "Rabbit Problem," among other examples, changed math as we know it. It is my sincere ambition that the book inspires an appreciation not only for the math but also for the man who propagated it: Leonardo Pisano, a.k.a Master Fibonacci.

Tarek I. Saab
Founder, fibonacci.com

PART I

I. ORIGINS OF THE FIBONACCI SEQUENCE

In the early thirteenth century, in the midst of a period of dramatic transformation, medieval Italy enjoyed significant commercial growth and economic stability. The country was awash in capital, which was used to finance the construction of breathtaking Gothic cathedrals and numerous universities. Much of that money came from taxes levied on profits from the extensive trading that occurred between countless merchants who traveled frequently to and from Mediterranean cities and ports to the west, and Asia Minor, Syria, and Baghdad to the east. Some intrepid adventurers even traveled so far as India and China, bringing back with them exotic supplies and novel ideas, beliefs, and methodologies. Exposure to people, languages and cultures in and from other parts of the world through trade and war gradually led to an intellectual transformation in Europe. In "Leonardo of Pisa and His Liber Quadratorum," American Professor R. B. McClenon explains that the crusades "awakened the European peoples out of their lethargy of previous centuries" and "brought them face to face with the more advanced intellectual development of the East." Marco Polo, McClenon reminds us, was only "the most famous among many who in those stirring days truly discovered new worlds."

In addition to being carriers of merchandise, traders and political representatives became "network specialists," comprising a "textual network" of sorts, providing a continuous exchange of written texts covering commercial, political, religious and literary topics (Roselaar). Courts which had access to unique intellectual resources reaped financial rewards; wealthy courts, therefore, financed some scholars permanently, sponsoring those who travelled from court to court, contributing to the spread of new ideas. Knowledge was also circulated via the translation and transcription of books that were usually copied at monasteries ("Transition"). Among the most important ideas brought into Europe this way was the Hindu numeral system.

HINDU NUMERALS ARE INTRODUCED TO EUROPE

Antonio Cifrondi, *Euclid*, 17th century, oil

Perhaps the most influential scholar of Western mathematics was the Greek mathematician Euclid, who lived from about 325 B.C. to about 265 B.C. His treatise on geometry, *The Elements*, written in thirteen books (chapters), contained everything known at that time about number theory. Notably, it includes an original geometric treatment of irrational numbers as well as the first known written definition of the "extreme and mean ratio" between the greater and lesser segments of a line, what is presently known as the Golden Ratio (Fitzpatrick). The significance of Euclid's book, *The Elements*, was evidenced by the fact that only the Bible was printed and studied more at that time ("Euclid"). Another important contributor to progress was the Salerno physician Constantine, who, early in the eleventh century, traveled thirty-nine years throughout Africa, Asia, and India, learning the Oriental sciences, and whose manuscripts were used as textbooks for centuries (Smith and Karpinski).

Hindu numeral forms appeared in Christian Europe long before manuscripts documented their arrival, for traders, travelers and ambassadors carried them from the East to various European markets. Most traders were rather good at transporting and applying new information and business methods but were less concerned with documenting sources, so pinpointing an exact date of adoption of Hindu numeral forms remains a mystery (Smith and Karpinski).

CALCULATING: FINGERS TO ABACUS

For centuries, traders in the Muslim world and Europe used either finger arithmetic or a mechanical abacus to perform calculations. The earliest of these devices were simple boards dusted with sand on which numbers could be traced. The Hebrew word for dust, avaq, may be the origin of the name "abacus." Later versions consisted of a flat board upon which were drawn or carved ruled lines; small pebbles were placed and moved around on these lines to indicate addition or subtraction. Since the Latin word for a pebble is calculus, this form of early calculation became known as "calculus" (Devlin, *Finding* 30). Medieval abaci had counters sliding along wires. Typically, an abacus had four wires, with beads on each wire representing units. An abacus was sufficient for conducting simple arithmetic operations, but users were at an enormous disadvantage when attempting to handle more complex computations. Since medieval merchants

Rechentisch (Holzschnitt vermutlich aus Straßburg) * Scan aus: Karl Menninger: "Zahlwort und Ziffer. Eine Kulturgeschichte der Zahl", Dritte Auflage, Göttingen 1979, S. 153

simply added and subtracted most of the time, they could manage using just finger calculations or Roman numerals. The fundamental disadvantage was, of course, the lack of a place-value system. (Livio 93-94).

Before man used written symbols for math, he used images to represent numerals and mathematical operations. Finger signs were particularly convenient and popular for calculating, but successful utilization required a great deal of practice to develop skill and dexterity. Nevertheless, because it was a reliable calculation method, the skill of calculating with fingers was preserved in various societies, passed from one generation to the next ("Calculating").

Finger calculations facilitated communication across language barriers for centuries. Ever desirous of educating the masses, scholars transcribed finger calculating representations in books; the earliest known transcription was written by the Venerable Bede (circa 673–735), an English Benedictine monk. In his book, *De Ratione Temporum*, he provides "a complete explanation of counting with fingers and rules for this method of calculating." Drawings in his text show that a person using both hands is able to represent all the numbers from 1 to 9,999 ("Calculating").

Calculating with fingers remained popular even after the introduction of Indo-Arabic numerals because it enabled higher calculations as well as counting, addition and subtraction. Scholars considered finger counting so important that a calculation manual was regarded as incomplete if it did not contain drawings showing the various finger positions ("Calculating").

CALCULATING: FINGERS AND HINDU-ARABIC NUMERALS

This latter fact explains why Leonardo Pisano's first book, *Liber Abaci*, includes instruction on "how the numbers must be held in the hands" (Devlin, *Finding* 89). At the end of the first chapter he includes a sophisticated system of finger counting as an aid for calculating in the "new" place-value system. The ancient method of finger counting was not deposed by the abacus, nor was it truly pushed into the background in Central Europe until the triumph of written calculation with Indo-Arabic numerals well into the thirteenth century ("Calculating").

Venerable Bede, De Temporum Ratione (725), Paris BN, MS Lat. 7418, f. 3v

The problem with the abacus or finger calculations is that these methods require considerable practice to achieve accuracy and speed. Additionally, there is no way to correct errors or check for accuracy because "neither method leaves a record of the calculation." When trading, one needs to be able to inspect records and audit transactions regularly, explains Keith Devlin, the mathematical historian who authored *Finding Fibonacci: The Quest to Rediscover the Forgotten Mathematical Genius Who Changed the World* (31). A better method was sorely needed.

In Baghdad, among the Arab scholars who studied and translated Greek and Hindu mathematical texts was a distinguished mathematician called Abu Abdallah Muhammad ibn Musa Al-Khwˆarizmˆı (ca. 780 to ca. 850). By the middle of the twelfth century, both of Al-Khwˆarizmˆı's books had been translated into Latin by scholars. These became essential resources for Europeans who wanted to learn the new mathematics. But Al-Khwˆarizmˆı's approach did not allow people to correct errors because it "involved cancelation and over-writing, which made it impossible to track the course of the calculation after it was completed" (Devlin, *Finding* 76-80).

Although Al-Khwˆarizmˆı wrote about the Hindu-Arabic numerals in the ninth century, translations of his work did not appear in Europe until the twelfth century. Yet, even then, the Hindu-Arabic numerals did not completely replace the daily and commercial use of Roman numerals until the fifteenth century. When a region adopted the Hindu-Arab numerals, they were often only used by mathematicians, surveyors and scientists; this was the case even in Arab lands (Devlin, *Man* 42).

European (descended from the West Arabic)	0	1	2	3	4	5	6	7	8	9
Arabic-Indic	٠	١	٢	٣	٤	٥	٦	٧	٨	٩
Eastern Arabic-Indic (Persian and Urdu)	۰	۱	۲	۳	۴	۵	۶	۷	۸	۹
Devanagari (Hindi)	०	१	२	३	४	५	६	७	८	९
Tamil		க	உ	ங	ச	ரு	சா	எ	அ	கூ

Evolution of Hindu-Arabic Numerals, Source: Wikimedia

Calculating with Hindu-Arabic numerals coexisted with the use of the abacus until the early thirteenth century. Indeed, one of the reasons people may have been reluctant to discard Roman numerals altogether is because they are "well suited for use on the abacus." Before he was made Pope Sylvester II in the tenth century, Gerbert of Aurillac was a scholar and teacher who invented a new abacus, known as the Gerbertian abacus. The counters of his abacus were marked with Hindu-Arabic numerals ("Transition").

Some twelfth-century translations of Arab treatises on algorism - the art of calculating by means of nine figures and zero (Merriam-Webster) - present calculations conducted with Roman numerals while others use Hindu-Arabic numerals. The early thirteenth century brought the appearance of some "very influential treatises on algorism" which "show a greater acquaintance with the new number system than the translations from the 12th century" ("Transition"). Yet, number forms varied widely from region to region, and nowhere in Europe were Hindu-Arabic numerals used exclusively.

Not long after Al-Khw^arizm^ı, the great Egyptian mathematician Abu Kamil (c. 850-c. 930), author of *The Book of Algebra,* sometimes worked on problems until he found all possible solutions; for one problem, he calculated and recorded 2676 solutions! (Sesiano). This unique, meticulous approach was adopted by Leonardo Pisano.

LEONARDO PISANO, A.K.A. FIBONACCI

Al-Khw^arizm^ı's text was written in the ninth century for use by merchants as well as scholars and astronomers (Devlin, *Man* 16). In the West, however, only educated men read the Latin translations of his works. Therefore, medieval Italian merchants who had not had the benefit of formal education were largely (if not completely) unaware of the Hindu-Arabic numerals and how to apply them for easier calculation. To remedy this situation, the Italian Leonardo Pisano described methods in his book *Liber Abaci* which were intended, from the outset, to be mastered by merchants and other businessmen as well as scholars (Devlin, *Finding* 80).

As was common practice at the time, Leonardo as an author borrowed ideas, methods, and explanations (sometimes the very words) of knowledgeable experts preceding him. Leonardo acknowledged that he was aware of Al-Khwˆarizmˆı's books and commentaries on them; he had possibly even studied them. He made "no claims of originality in *Liber Abaci,* although he did so in another of his books, *Liber Quadratorum*" (Devlin, *Finding* 81).

Some people confuse Leonardo of Pisa with Leonardo da Vinci, polymath of the Renaissance, but the famous inventor, scientist, and painter of the "Mona Lisa" was born in Vinci, between and Florence, in 1452, about 200 years after the death of Leonardo of Pisa.

All that we know for certain about Leonardo Pisano is contained in a few sentences he wrote about himself in the 1228 edition of his famous *Liber Abaci* (Horadam). Presented here is the paragraph in English, translated from Latin by Richard E. Grimm in "The Autobiography of Leonardo Pisano" (1973):

Unknown, Derived from an engraving appearing in *I benefattori dell'umanità;* vol. VI, Firenze, Ducci, 1850.

"After my father's appointment by his homeland as state official in the customs house of Bugia for the Pisan merchants who thronged to it, he took charge; and, in view of its future usefulness and convenience, had me in my boyhood come to him and there wanted me to devote myself to and be instructed in the study of calculation for some days. There, following my introduction, as a consequence of marvelous instruction in the art, to the nine digits of the Hindus, the knowledge of the art very much appealed to me before all others, and for it I realized that all its aspects were studied in Egypt, Syria, Greece, Sicily, and Provence, with their varying methods; and at these places thereafter, while on business, I pursued my study in depth and learned the give-and-take of disputation. But all this even, and the algorism, as well as the art of Pythagoras I considered as almost a mistake in respect to the method of the Hindus. Therefore, embracing more stringently that method of the Hindus, and taking stricter pains in its study, while adding certain things from my own understanding and inserting also certain things from the niceties of Euclid's geometric art, I have striven to compose this book in its entirety as understandably as I could, dividing it into fifteen chapters. Almost everything which I have introduced I have displayed with exact proof, in order that those further seeking this knowledge, with its pre-eminent method, might be instructed, and further, in order that the Latin people might not be discovered to be without it,

as they have been up to now. If I have perchance omitted anything more or less proper or necessary, I beg indulgence, since there is no one who is blameless and utterly provident in all things."

This exceptionally brief glimpse of the man's life provides provocative clues about "his personality and the mathematical quality of his mind." These few sentences reveal that he was a man with great intellectual curiosity who was excited by mathematical scholarship and valued it so much that he wished to share his knowledge with the working class, not just scholars. The words with which he concludes his introduction to *Liber Abaci* show that he is open-minded, willing and even eager to learn new ideas, receptive to constructive criticism, and modest despite his genius. He was not only revered by readers, students, and authorities of his day, but modern scholars also conclude that he was a man well worthy of respect; some even develop a "warmth of feeling for the modest humility of the man" (Horadam).

In increasing numbers, serious mathematics scholars today are beginning to recognize that Leonardo Pisano may have been the "most noteworthy mathematical genius of the Middle Ages," and definitely the most influential of all medieval writers in promoting the Hindu-Arabic numerals to European scholars. (Smith and Karpinski).

His given name was Leonardo, so his full name was Leonardo of Pisa, or Leonardo Pisano in Italian, and he was born about the year 1175. He is better known by the nickname that apparently was given him by the math historian Guillaume Libri in 1838, six-and-a-half centuries after his birth: Fibonacci [pronounced fib-on-arch-ee] which is short for filius Bonacci, is the name Leonardo ascribed himself in *Liber Abaci* (Livio; Knott).

An abbreviation of the Latin phrase, "filius Bonacci," Fibonacci means "the son of Bonaccio." Fi'-Bonacci, then, is a general name like the English names of Robin-son and John-son or Bronson (Brown's son) (Knott, Smith and Karpinski). But (in Italian) Bonacci is also the plural of Bonaccio; therefore, two early writers on Fibonacci (Boncompagni and Milanesi) regard Bonacci as his family name (as in "the Smiths" for the family of John Smith).

Fibonacci referred to himself in writings as "Bonacci" and "Bonaccii" as well as "Bonacij." It was common in medieval Pisa to mix spoken Italian with written Latin; there remains uncertainty about which is the most correct spelling of what may be considered his surname. However, it is certain he did not use "Fibonacci" when referring to himself (Knott).

Bonacci may be a kind of nickname meaning "lucky son" (literally, "son of good fortune") which certainly seems representative of the fortuitous circumstances of his life, what little is known of it (Knott). Finally, Mario Livio explains in *The Golden Ratio* that the nickname Fibonacci can also mean “son of good nature” (Livio 92). Given that he was called “beloved” by his countrymen in an official proclamation in the year 1241, this last explanation for the meaning of his name, “Bonacci,” may be most plausible of all. Leonardo’s humility is evident in the introduction to *Liber Abaci,* when he says, “If by chance I have omitted anything more or less proper or necessary, I beg forgiveness, since there is no one who is without fault and circumspect in all matters” (Livio 94).

Occasionally he signed his name as Leonardo Bigollo since he was one who had studied in foreign lands and, in Tuscany, bigollo means “a traveler” (O’Connor and Robertson; Smith and Karpinski). Interestingly, the term may also mean “good-for-nothing” (O’Connor and Robertson) or “man of no importance” (in the Venetian dialect) (Livio 93). These are similar to yet another meaning of the word, which is “blockhead,” so some have thought Leonardo might have adopted the term as a humble way to identify himself or that it “may have been applied to him by the commercial world or possibly by the university circle,” and subsequently taken upon himself as a challenge “to prove what a blockhead could do” (Smith and Karpinski).

The historian of mathematics, Guillaume Libri, appears to have been the first to use the name “Fibonacci” when referring to Leonardo; this was found in a footnote in his 1838 book *Histoire des Sciences Mathematique en Italie* (*History of the Mathematical Sciences in Italy*) (Livio 93). Since most modern authors now call Leonardo Pisano “Fibonacci,” one should be prepared to see any of the above variations of his name when reading about him.

Contemporary writers identify Leonardo's father as Guglielmo Bonacci (William), and mention that he was a kind of customs officer in the present-day Algerian port city of Béjaïa, formerly known as Bugia or Bougie, where wax candles (still called "bougies" in French) were exported to France (Devlin, *Man* 9). A legal document provides the name of one brother, Bonaccinghus, but nothing further is known about the rest of Leonardo’s family (Livio; Smith and Karpinski).

Leonardo says his father was a kind of "state official;" it is unclear what the position entails precisely. By some accounts, Guglielmo was elected the consul for Pisa. This was a highly prestigious public office so, if this was in fact the case, most likely William was a wealthy merchant himself before he became the representative for the Pisan merchants who were trading in Bugia (O'Connor and Robertson).

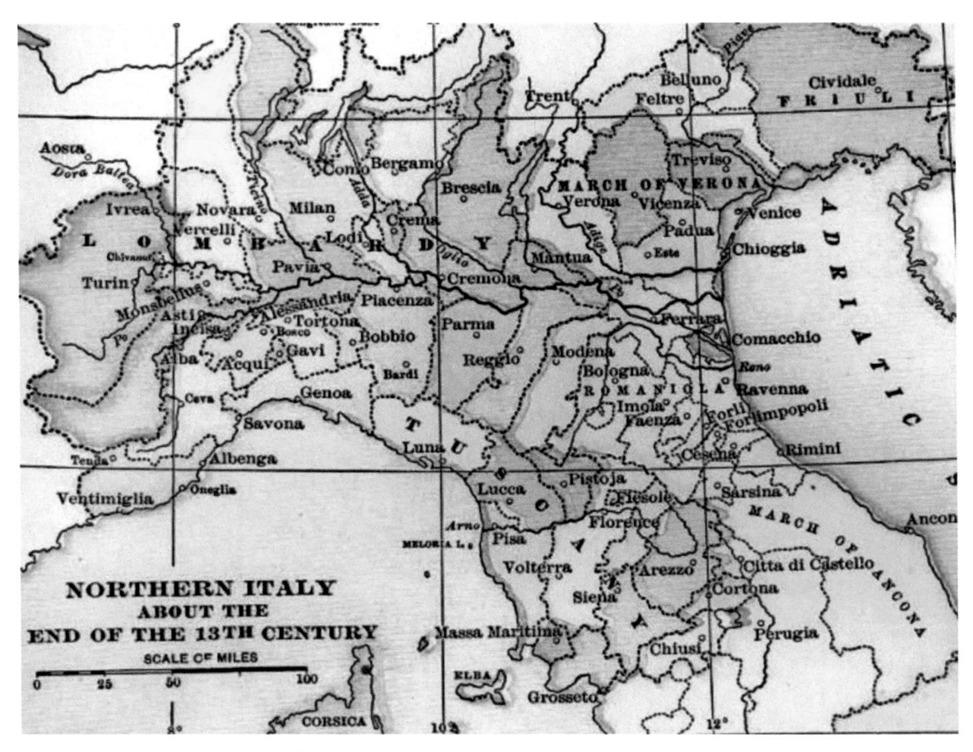

Source: Dow, Earle W. Atlas of European History. New York: Henry Holt and Company, 1907.

Pisans exported European goods to North Africa through Bugia and brought various Eastern luxury items to Europe, including silks, spices, and a fine grade of beeswax useful for candles, and high-quality leather (Devlin, *Man* 39). At the close of the twelfth century, Bugia was a center of African commerce sheltered by Mount Gouraya at the mouth of the Wadi Soummam near Cape Carbon. Being a commercial agent of one of the most important Islamic ports on the Barbary Coast concerned with taxation of trade between Pisa and North Africa, Fibonacci's father would have spent a lot of time monitoring trade transactions in the bustling harbor. He must have been fluent in Arabic since he was expected to maintain relations with the Muslim authorities, safeguard the rights of the fondaco (trading post), keep records of the goods passing through, and oversee the proper levying of taxes (Devlin, *Man* 41). Of course, Guglielmo would have recognized the value of accounting skills and surely believed "the future would be bright for people who understood numbers thoroughly" ("Fibonacci" *Famous*). Thus, Leonardo was summoned by his father and taken as a child (Devlin says he was likely no more than fifteen years old) to the Arab port city on the North African coast for the furtherance of his mathematics (accounting) education under the tutelage of Moorish masters (Devlin, *Finding* 13; Livio; Smith and Karpinski).

LEONARDO'S HOME: PISA

Pisa at the dawn of the thirteenth century was in the midst of a "golden age" of commercial, religious, and intellectual prosperity (Smith and Karpinski). She was a busy port and a major Mediterranean trading hub for the importing and exporting of merchandise from both inland and overseas (Livio 93).

Leaning Tower of Pisa

"When Leonardo was growing up, a new, heavily fortified city wall was being constructed, to protect the city from attack both by Muslims – this was the time of the Crusades – and by rival Italian cities as part of the ongoing political struggle" between the empire and the papacy (Devlin, *Finding* 57). Rather than having a detrimental effect, this interurban warfare contributed to the stimulation of commercial activity (Smith and Karpinski). In addition, the city's many naval fleet victories secured the profitable expansion of trading territory throughout the Mediterranean and also into Syria, significantly enriching the government coffers. At this time, Pisa's military and economic dominance in the Mediterranean rivaled those of Genoa and Venice ("Editors").

In addition to city walls, the thriving economy inspired the people of Pisa to begin building an impressive cathedral complex, the Field of Miracles, comprising a cathedral, a baptistery, a bell tower and a cemetery. The 179-foot bell tower, begun in 1173, is known as the Leaning Tower of Pisa and is universally recognized as an unofficial emblem of Italy.

Leonardo would have encountered a commercial frenzy every time he accompanied his merchant father to the customs houses or in bustling streets beside the crowded River Arno. Such activity required unceasing measurements of merchandise quantities and price negotiations; as a customs officer, Leonardo's father would need to calculate import tax levies and audit ships' manifests. He must have himself been just one of the many scribes and stewards recording inventories, orders and transactions. Prices were recorded in librae (pounds), solidi (shillings), and denarii (pennies); scribes entered the values in long columns, using Roman numerals, and used abaci to perform the calculations (Devlin, *Finding* 62).

Since Leonardo's father was a prosperous merchant participating in all of these business activities, the enormous power and indispensability of arithmetic surely made an impression upon the young boy (Devlin, *Finding* 58). At the wharves, Leonardo observed other professions besides merchants and traders. He saw surveyors and engineers and shipbuilders working with math. Pisa built and maintained a fleet of hundreds of naval and commercial ships, so incoming cargo in the harbor consisted often of timbers for building as well as sacks of grain, salt from Sardinia, squirrel skins from Sicily, and goatskins from North Africa. There were shipments of leather, alum, and dyes for the textile manufacturers of Italy and northwest Europe. Crates of spices sailed in from the Far East and barrels of wine were common. Among her many exports, Pisan ships regularly transported "barrels of Tuscan wine and oil, bales of hemp and flax, and bars of smelted iron and silver" (Devlin, *Finding* 59).

Presumably, Leonardo had on countless occasions watched scribes as they listed prices in Roman numerals and added them up using an abacus. The impracticality of the Roman system for complex applications in commerce and trade (such as currency conversion, or commission calculations) would have been frustratingly apparent to an inquisitive merchant's son who surely wondered whether there was not a more efficient way of calculating than Roman numerals and an abacus (Devlin, *Finding* 62 and *Man* 15).

EDUCATION AND TRAVEL

Medieval churches provided formal education for wealthy citizens, so Leonardo would likely have attended school "between ten and twelve years of age in the cathedral in Pisa" (Devlin, *Finding* 37). Dr. Thomas O'Shea, scholar, author, and retired educator in British Columbia, shares an excerpt from a memoir written by a gentleman of Pisa born in 1308 which "gives us some idea of how students of that time were prepared for their future in the world of trade and commerce." It is reasonable to expect Leonardo's education would have been similar. Olinto Bernardini writes:

"Naturally, it was assumed that my two brothers and I would each assume a place in the family Company. It was also assumed that we would become competent in the mathematical skills used to run businesses in our parts. Therefore, at the age of ten, as was the custom, my father sent me to begin the study of abaco. This course, under Maestro Pietro Cataneo of Pisa, lasted two years, but well before it was over, it had changed my life forever.

Abaco schooling normally began with an introduction to the Hindu numbers, and with an explanation of the place value that gave them meaning. But already familiar with this numeration, and well-versed in multiplication and addition facts, I quickly advanced to the next four mute [stages] where students were taught division and fractions. All this I devoured eagerly and before long had even caught up to students who were one year older. From there, the maestro introduced me to the core of abaco learning: the mathematics of business that even my great-grandfather had used in his quest to make a fortune.

At this level, that is, in the sixth and seventh mute, we were taught to work with prices, barter, partnerships, alligation, proportions, monetary systems, measurements, interest and discount. Our teacher took great care in imparting this knowledge to us, for he appreciated that he had but two years to prepare us for important positions in business."

The two-thousand-mile (ca. three-thousand kilometer) journey to Bugia would have taken approximately two months, during which time the ship would have stayed close to land for greater protection from the weather and to pull into ports for trading. In these ports Leonardo would have met Arab traders who had ventured in their travel even farther afield than the Italians, "journeying not only around the Mediterranean but to Russia, India, and China, and deep into the interior of Africa" (Devlin, *Man* 41). Arriving in Bugia, Leonardo would have likely joined his father in the sizable Italian community near the harbor (Devlin, *Man* 41).

Later, in his autobiographical paragraph in *Liber Abaci* (1228), Leonardo explained that, after leaving Bugia, he travelled extensively around the Mediterranean Sea, touring the cities of the East along the Mediterranean coast, visiting the great markets of Egypt and Asia Minor, Syria, Sicily and Provence, Constantinople and Greece. While he was touring, he met and learned from scholars as well as from merchants, "imbibing a knowledge of the various systems of numbers in use in the centers of trade" (Smith and Karpinski). With a mastery of Arabic, Leonardo would have been able to broaden his mathematical knowledge well beyond what he had been able to observe in the Bugian marketplace (Devlin, *Man* 46). He wrote:

"When my father, who had been appointed by his country as public notary in the customs at Bugia acting for the Pisan merchants going there, was in charge, he summoned me to him while I was still a child, and having an eye to usefulness and future convenience, desired me to stay there and receive instruction in the school of accounting. There, when I had been introduced to the art of the Indians' nine symbols through remarkable teaching, knowledge of the art very soon pleased me above all else and I came to understand it…"
(O'Connor and Robertson)

This philomath inevitably realized that the mathematics system used by Oriental merchants had many advantages over all others. In fact, he admitted (later in the paragraph) that he considered every other mathematical system as almost a mistake compared to the method of the Hindus.

Leonardo sought instruction in the Hindu-Arabic arithmetic system and practiced it carefully. He recognized its superiority over the clumsy Roman numeral system in the West, and accordingly decided to write a book to explain the superior system and its applications to the Italians (McClenon). Ending his travels around the year 1200, the scholar returned to Pisa and proceeded to share the "treasure of knowledge" he had acquired, supplemented by ideas of his own (O'Connor and Robertson; "Biography"). However, Leonardo did not merely copy the works of others; he was a brilliant mathematician in his own right. He was exceptionally skillful at explaining mathematical theories, problems, and solutions in a way that the common reader could understand.

LIBER ABACI 1202 (1228)

Leonardo's greatest intellectual legacy is his book, *Liber Abaci* (*The Book of Calculation*). This book provides almost the only biographical information we have about him. No one knows why Leonardo dedicated the book to Michael Scotus, who was the court astrologer to Holy Roman Emperor Frederick II, nor why he included the autobiographical information (Devlin, *Man* 43). It was uncommon for mathematically-focused texts to include such information. Mathematician/ historian Devlin explains this is because "most mathematicians are interested in mathematical results, not the people who discover them" (*Finding* 68). He says, "Mathematical truth is completely independent of human judgment, immutable, eternal, forms its own abstract world," so, "to the mathematician, the historical details are of little relevance; it doesn't matter when someone first proved a theorem; the focus is on the development of the ideas and how one train of thought led to another" (*Finding* 69). The first edition of *Liber Abaci* was a dense work suited more to scholars than the average man. But Leonardo seemed very occupied with producing practical solutions to common problems. Thus, in the preface of the second edition he revealed something of the intellectual heritage which inspired him to write the book in the first place.

The first copy of *Liber Abaci* appeared in Pisa in 1202. "When he finished writing it, he would have taken it to a local monastery to have copies made by the monks." This is such a laborious method of publication that it could have taken a year or more to copy a manuscript as long as *Liber Abaci* (over 400 pages). After receiving peer commentary and suggestions for improvement, "he made changes and added to its contents, culminating in the second edition published in 1228" (Devlin, *Finding* 85-86).

In modern publications, the most common spelling of the book's title uses just one letter "b" in the word "Abaci," a declension of the word 'abacus' in medieval Italy. This spelling is nonsensical, though, says Devlin, because "Leonardo was in fact showing how to do arithmetic without the need for any such device as an abacus." Leonardo himself used the spelling "Abbaci" to mean 'calculation.' The words are similar in spelling, but very different in meaning. The first known written use of the word abbacus with this spelling was, in fact, in the prologue of his book, an intentional spelling change to differentiate Leonardo's new method. It must have caught on, for in the years after Leonardo, the word abbaco was widely used in medieval Italy to describe the practice of calculating with the Hindu-Arabic number system (Devlin, *Finding* 11).

Since medieval authors rarely gave their manuscripts a title, neither did Leonardo; the name for his book comes from his opening statement: "Here begins the Book of Calculation; Composed by Leonardo Pisano, Family Bonacci; In the Year 1202."

Original manuscript, 1228, Leonardo da Pisa, Liber abaci, Ms. Biblioteca Nazionale di Firenze, Codice magliabechiano cs cl, 2626, fol. 124r Source: Heinz Lüneburg,

In later writings, he also referred to the work as *Liber Numerorum,* and in the dedicatory letter for his book *Flos* he referred to it as his *Liber Maior de Numero* (Devlin, *Man* 12).

Many consider Leonardo's book the greatest arithmetic text of the Middle Ages, for he was the first mathematician to demonstrate the superiority of the Hindu-Arabic numeral system versus the Roman system exemplified by Boethius; he did this by providing numerous examples of how to solve problems related to every major contemporary field of business. It is true that Leonardo's *Liber Abaci* was not the first book written in Italy to introduce and explain the Hindu-Arabic numerals, but no work previously produced was comparable in value, either in content or in quality of the exposition (McClenon).

Leonardo consulted many sources to write *Liber Abaci,* primarily Arabic texts or Latin translations thereof. He undoubtedly included information gleaned from his many discussions with the Arab mathematicians he encountered on his travels. He provided rigorous proofs to justify the methods, in the fashion of the ancient Greeks. For example, Abu Kamil's book has seventy-four worked-out problems, and many of the more complicated ones, with identical solutions, are found in *Liber Abaci* (Devlin, *Man* 47, 57, 61). Leonardo explains the mathematical reasoning for each problem extensively, providing numerous examples and variations which were most valuable to the merchants and laymen learning the new calculating system (Devlin, *Finding* 80).

Liber Abaci is a general book of mathematics, but it differs from most because the author's main purpose was to encourage everyone (especially merchants) to abandon Roman numerals and use the superior Indian system of numbers. Leonardo was an advocate for systemic change. Knowing the superiority of the new system for business, he devoted several chapters of his book to showing calculations of profit, interest, and currency conversions. Some scholars think the book was "too advanced for the mercantile class, and too novel for the conservative university circles" which were resistant to adopting Arabic numerals (Smith and Karpinski; Radford). Furthermore, it was so comprehensive, it has been called "encyclopaedic" ("Biography"). Moreover, at the time he published his book, only a few people knew the Hindu-Arabic numerals we use today: privileged European intellectuals who cared to study the translation of the works of Al-Khwˆarizmˆı and Abu Kamil. Nevertheless, because Leonardo was so convinced that the Hindu-Arabic numerals and the place-value principle were far superior to all other methods, he devoted the entire first seven chapters of his book to explanations of Hindu-Arabic notation and its use in practical applications (Livio 92).

The book is relevant for mathematicians today because of "the mathematical insight and originality of the author, which constantly awaken our admiration, and also on account of the concrete problems" (McClenon). It is also of interest to historians, sociologists and economists, because it provides much information about the society in which he lived. For example, through his book, we learn that Pisan ships transported "pepper, a very important item of merchandise, and that the Pisan colony in Constantinople traded extensively with Egypt. Further evidence is also gleaned about the relative values of money coined in the mints of different cities, and about the problem of alloying of coins to be minted" (Horadam).

In Pisa, Leonardo not only became an author of books, but he was likely also a maestri d' abaco or "teacher of business arithmetic." As such, he became so famous that the Roman emperor asked the mathematics expositor to give a public demonstration of his ability during one of the emperor's visits to Pisa (Livio 95).

Holy Roman Emperor Frederick II (1196-1250) was called Stuper Mundi ("Wonder of the World") by his contemporaries because he was a highly educated and inquisitive man who "encouraged learning and scholarship of every kind;" he even conducted scientific experiments and wrote books of his own. Having a special interest in mathematics and impressed after reading *Liber Abaci,* the emperor invited Leonardo to his palazzo in Pisa (Horadam).

One of the emperor's court mathematicians, Johannes (John) of Palermo, proposed three mathematical problems for Fibonacci to solve. Leonardo had been provided the problems in advance but the ingenious way he solved and explained the solutions was astounding (Horadam).

The Medieval Italian Cultural Association claims Emperor Federico II was so impressed that he granted Leonardo an annuity which enabled him to devote himself to his studies ("Arabic Numerals"). If this is true, that could explain how Leonardo was able financially to devote more time to writing a revised version of *Liber Abaci*. Indeed, soon after meeting the emperor, the mathematician dedicated his next major work to the emperor, perhaps in appreciation for sponsorship; he also published the solutions he had presented to the imperial court: two in Flos (a copy of which he sent to Frederick) and one in *Liber Quadratorum* (Devlin, *Man* 90-92).

LIST OF LEONARDO PISANO'S MATHEMATICAL WRITINGS

1202 (1228): *Liber Abaci* (*The Book of Calculation*)

1220: *Practica Geometriae* (*The Practice of Geometry*), a mixture of pure mathematics, theorems, proofs, and practical applications of geometry, such as using similar triangles to calculate the height of tall objects

Before 1225: "Epistola" and "Magistrum Theodorum" (A Letter to Master Theodore), a letter to Frederick II's philosopher Theodorus Physicus, solving three problems in mathematics

1225: *Liber Quadratorum* (*The Book of Square Numbers*), a highly mathematical number theory book dealing with solutions to Diophantine equations

1225: *Flos* (*The Flower*), solutions to problems in algebra

n.d. (no date known): *Di Liber Minor Guisa* (*A Smaller Manner*), a book on commercial arithmetic (No copies exist today.)

n.d.: *Commentary on Book X of Euclid's Elements* (No copies exist today.)

(Horadam; "Education")

Practica Geometriae (1220)

The *Practica Geometriae* (*The Practice of Geometry*) is a substantial work on geometric practice (surveying, area and volume formulas for plane figures and bodies); it also contains a wide variety of interesting theorems which represent "a considerable advance over the Geometry of Boethius and Gerbert (Pope Sylvester II)" (Horadam). Leonardo well understood Euclidean geometry and the mathematical methods demonstrated in *Practica Geometriae* reproduce the brilliant techniques found in the works of others, particularly Abu Kamil's *On the Pentagon and the Decagon* (Livio 96). Though it "shows no such originality as to enable us to rank Leonardo among the great geometers of history, it is excellently written, and the rigor and elegance of the proofs are deserving of high praise" (McClenon). Furthermore, *Flos*, *Magistrum Theodorum*, and the *Liber Quadratorum* are so original and instructive, they show well the remarkable genius of this brilliant mathematician.

Liber Quadratorum (The Book of Squares) (1225)

Leonardo describes his presentation to Emperor Frederick II at the Pisan court in the dedication to this book, *Liber Quadratorum*, dated 1225 (McClenon). In it, he demonstrates a virtuosic command of number theory. The book, among other things, examines methods to find Pythogorean triples (O'Connor and Robertson). Even more impressive are his presentations of the properties of the squares and tasks that lead to quadratic equations; it has been called the most important work of number theory between Diophantus and Fermat (Gies; "Biography").

Liber Minoris Guise (n.d.)

The *Liber Minoris Guise* (*Book in a Smaller Manner*) is a manuscript on commercial arithmetic referred to with description (rather than by title) by the Pisan several times as he was comparing it to his more extensive *Liber Abaci*. Additional proof this missing work once existed is a reference to it by an abbacus author who refers to Leonardo's *Libro di Minor Guise o Libro di Merchanti* (*Book in a Smaller Manner or Book for Merchants*) (Devlin, *Finding* 28).

Because Leonardo lived two hundred years before mechanized printing was widely available in Europe, his books were handwritten and the only way to have a copy was to have a scribe handwrite another copy. Of his books, copies remain of *Liber Abaci* (1228), *Practica Geometriae*, *Flos*, and *Liber Quadratorum*. Regrettably, another of his manuscripts now completely lost is his commentary on Book X of Euclid's *Elements* (O'Connor and Robertson). No copies of his first *Liber Abaci* exist, and of his 1228 revision, only fourteen copies have been found. Seven are complete or nearly so while seven are fragments, consisting of between one-and-a-half and three of the book's fifteen chapters (Devlin, *Finding* 82).

LEONARDO'S DEATH AND MEMORIAL

No one knows when or how Leonardo of Pisa died (Horadam). After publication of his revised version of *Liber Abaci* in 1228, only one known document refers to him. This is a decree made by the Republic of Pisa in 1241 in which "the serious and learned Master Leonardo Bigollo" was granted an annual honorarium of twenty Pisan pounds plus expenses for services to the city. Historians believe this was either in return for advising on matters of accounting (such as advising on financial contracts stipulating long-term debt commitments to creditors) and teaching the citizens, or for service as city auditor (O'Connor and Robertson; Devlin, *Man* 99).

Even though he was famous in his lifetime as a brilliant mathematics expositor and, later, as a respected public servant, Leonardo was forgotten within 200 years of his death. Devlin says this should not be surprising since, "other than nobility, few people had anything recorded about them, even those who had achieved great things" (*Finding* 22). His name did not appear in any book on the history of science or mathematics for 400 years.

In 1495, Luca Paciolo resurrected the name of Leonardo Pisano, more than 250 years after the Pisan decree (the last recorded proof that Leonardo was still alive). Pacioli printed a highly regarded, scholarly book titled, *Summa de Arithmetica, Geometria, Proportioni et Proportionalita* (*All That is Known About Arithmetic, Geometry, Proportions, and Proportionality*). The Venetian deliberately mentioned Leonardo as his most valuable source, stating, "Since we follow for the most part Leonardo Pisano, I intend to clarify now that any enunciation mentioned without the name of the author is to be attributed to Leonardo" (Devlin, *Man* 7). Despite this extraordinary endorsement, Leonardo's contributions to mathematic intelligence lingered in obscurity and "his influence languished for many centuries and indeed Mathematics made no real progress for 300 years" (Horadam).

Then, late in the eighteenth century, another Italian mathematician named Pietro Cossali (1748-1815) came across this single reference to Leonardo in Pacioli's book. Wondering why Pacioli was famous while the man whose work he "followed" was unknown, Cossali began to look for Pisano's manuscripts (Devlin, *Man* 8).

In 1838, French historian Guillaume Libri gave Leonardo the manufactured surname 'Fibonacci.' Then, in the 1870s, another Frenchman, the mathematician

Edouard Lucas, assigned the name "Fibonacci sequence" to a fascinating number sequence that surfaces when one tries to solve one of the more recreational problems in *Liber Abaci* (Devlin, *Finding* 24). Leonardo has been called Fibonacci ever since.

A few memorials commemorate Fibonacci's contributions to Italy, among them two street names - the quayside Lungarno Fibonacci (Fibonacci Way) in Pisa and the Via Fibonacci in Florence - and a statue of him (as seen on this book's cover) with a "kind, scholarly expression," in scholar's garb, in the Camposanto, a historical cemetery on the Piazza dei Miracoli in Pisa (Devlin, *Finding* 36, 49).

FIBONACCI'S INTELLECTUAL LEGACY

If what followed Leonardo 'Fibonacci' Pisano was an economic revolution, then *Liber Abaci* was the "gunshot that started the revolution." There has been discovered in many Italian archives thousands of medieval manuscripts of short, handwritten textbooks in "practical arithmetic" which are so numerous they form their own extensive genre. These manuscripts are called *libri d'abbaco* ("abbacus books") or *trattati d'abbaco* ("abbacus tracts"). "They are written in vernacular Italian, usually in the local dialect of the author. ...The earliest of these were handwritten but after the invention of the printing press in the fifteenth century they became recognizable as a genre and some became best sellers." (Devlin, *Finding* 25-27) Today there remains more than 400 such texts stretching over 300 years. Part of their significance lies in the quantity; they are proof that their "rapid proliferation" signifies "the importance people attached to learning the new arithmetic." It is evident that each one was written for a local audience, because the problems they presented were usually expressed in terms of the currency, weights and measures of the local town or region. Some were evidently written by more learned scholars who may well have been teachers who wrote them "to use in classes on practical arithmetic." Indeed, by the end of the thirteenth century, there were "a number of 'abbacus schools' (*scuole d'abbaco* or *botteghe d'abbaco*) where 'abbacus teachers' (*maestri d'abbaco*) taught practical arithmetic" (Devlin, *Finding* 25-27).

Scholars studying these manuscripts extensively have concluded that it is possible to "construct an ancestral tree of books" leading to the original document, or source, of all the others. Devlin explains, "The entire genre began with a single

'Abbacus Eve,' the mother of all abbacus books" and that book, he asserts, is *Liber Abaci*. He says we know the "Abbacus Eve" was "written by Leonardo himself because no other contemporary mathematician was as accomplished (or he would surely have left his own collection of writings)" (Devlin, *Finding* 28).

The Hindu-Arabic numerals Fibonacci championed were obviously of inestimable worth to the expanding commercial enterprises of medieval European society. "Of greater importance was the long-range impact on Science and Mathematics of the new system of numeration which he publicized" (Horadam). Nevertheless, few mathematicians over the centuries were aware of his brilliance, most likely because his texts were written in Latin and have remained untranslated into modern languages for so long (eight hundred years!) (Horadam).

Ironically, Fibonacci is known primarily because of the sequence bearing his name "but which he treated only lightly." Modern mathematicians have named an Association, a Journal, and a Bibliographical and Research Centre after Fibonacci, ensuring that his name (at least one of them) will not be quickly forgotten again (Horadam; Devlin, *Finding* 37).

PART II

II. THE FIBONACCI SEQUENCE

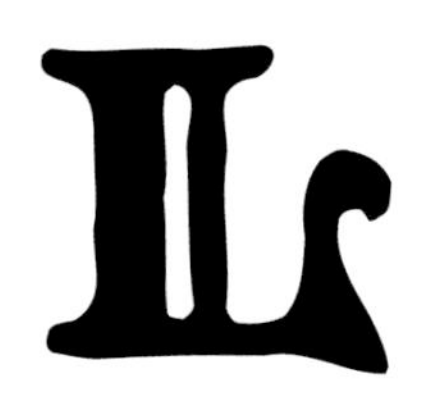

eonardo Pisano, or 'Fibonacci,' was a self-professed student of the arts of Greek theologian and mathematician Pythagoras (c. 569 - 475 BCE), who coined the term 'mathematics' (Μάθημα, ατος, το, that which is learned) to represent that abstract science which studies shape, quantity, and space (Donnegan 26). Pythagoras was primarily interested in number theories and their application to music rather than the use of numbers for everyday computation (O'Shea). Greek mathematician Euclid also greatly influenced Fibonacci; as highlighted previously, his thirteen books (chapters) on geometry in *Elements* (c. 300 BCE) provided definitions, postulates and axioms of geometry which Fibonacci knew well. *Elements* is considered by many the most scientifically significant mathematical work until the 20th century ("Euclid"). "Even today a large part of mathematical and geometrical elementary education is based on the Euclidean tradition" ("Euclid").

Long before Pythagoras or Euclid, man recorded counting by "scratching tally marks on a stick or bone" (Devlin, *Man* 13). Mathematics has evolved greatly since then; today, math equations are so inconspicuously calculated by computers that most people tend to think about mathematics "only in the day-to-day context in which they themselves are immersed" (Radford).

One of the earliest proofs of math skills practice is a papyrus written by an Egyptian scribe named Ahmes (~1650 BCE), who recorded a "series of 87 exercises and problems, presumably for students to try with the assistance and guidance of a teacher" (Levy 21).

The concept of "nothing" in math may have been represented first by a dot to indicate an empty placeholder, but zero was first used as a number in the seventh century (rather than as a mere concept) by the Indian mathematician and astronomer Brahmagupta, who also devised rules for its use as a number (Levy 93). Because the idea of nothing was important in early Indian religion and philosophy, it was much more natural for them to adopt a symbol for it than it was for the Latin (Roman) and Greek systems (Knott, "Brief"). Thus, zero was used in India for addition, subtraction, and multiplication but not division; the concept of dividing something by "nothing" was too difficult for even the brilliant Brahmagupta (Levy 93).

Islamic mathematicians in Egypt, such as Abu Kamil (c. 850 – c. 930 CE), produced important but "only incremental progress" in the development of algebra, particularly of the use of the Golden Ratio (Sesiano). Such incremental advancement may not have been revolutionary, but it was necessary for the preparation of later mathematicians (like Fibonacci) to push forward the next major math breakthrough (Livio 91).

Suleimān the Merchant, a well-known Arab trader of the ninth century, may have introduced the Hindu-Arabic symbols (including the numeral zero) to European markets, and "Abū 'l-Ḥasan ʿAlī al-Masʿūdī (d. 956 CE) of Baghdad traveled to the China Sea on the east, as far south as Zanzibar, and to the Atlantic on the west; he speaks of the nine figures with which the Hindus reckoned (Smith and Karpinski). Thus, Islamic mathematicians may have learned the number zero from India but "failed to make use of it in algebra." Hundreds of years later, Fibonacci did not consider zero a number like other numerals, either; instead, he referred to it as a symbol in his book *Liber Abaci* (Levy 93).

Educated and (of course) fluent in Latin, Fibonacci studied al-Khwarizmi's compendium of rules for calculating Hindu-Arabic arithmetic in a book which was later translated into Latin and given the title, *Algoritmi de Numero Indorum* (*Concerning the Hindu Art of Reckoning*) (Devlin, *Man* 24). First exposed to this book either in Bugia or perhaps while traveling the Mediterranean, it greatly influenced Fibonacci's understanding and practice of the Hindu-Arabic arithmetic.

USSR stamp of Muhammad ibn Musa al-Khwarizmi, 1983

Although Kamil and al-Khwarizmi were accomplished

mathematicians in the Arabic world and some in Europe were aware of their arithmetic strategies, a commercial revolution did not emerge from Baghdad at that time because Western commercialism "had not yet developed sufficiently for the new methods to have widespread impact" (Devlin, *Finding* 32).

In fact, early in the twelfth century, other books explaining the Hindu art of reckoning were written but the new numerals were not enthusiastically embraced. Slowly, however, Italian merchants and bankers who initially opposed the unfamiliar numerals and the new calculation methods eventually understood its advantages over the traditional method of using Roman numerals. Transitioning to the new math, for example, eliminated the need of counting boards and other primitive means of commerce and banking. Among these was the primitive use of tally sticks; the money value of a loan was written upon a tally stick which was split in two. The lender kept the biggest piece - the stock - becoming the "stockholder" (Seife 81).

Society was reluctant to adopt the Hindu-Arabic arithmetic system for many reasons, only a few of which will be mentioned here. Perhaps the most significant is the natural human aversion to change. Roman numerals had worked well enough with ancient counting devices and abaci for millennia; they had met the need for addition, subtraction, and multiplication. Moreover, little explanation was required to successfully operate an abacus (extensive practice was needed to truly excel in its use, however, especially when multiplying different orders of numbers) (Smith and Karpinski).

Another reason is that social discord between abacists (the advocates of the abacus) and algorists (those who favored the use of the Hindu-Arabic numerals) kept the newer, more efficient system from becoming universally adopted by European society for years. In fact, "merchants' ledgers featured Roman numerals throughout the Middle Ages, indicating that [some] remained staunchly abaci" (Levy 117).

Newly-established universities were sometimes antagonistic toward algorism but a more powerful impediment to the dissemination of the Hindu-Arabic math method was civil authority. At the end of the 13th century, the use of new numbers was forbidden by local governments in several Italian cities. Florence in 1299, for example, passed statutes forbidding moneychangers' guilds (bankers) from using Arabic numerals (Levy 117). Similarly, the statutes of the University of Padua required stationers to keep the price lists of books "non per cifras, sed per literas claros" (loosely translated, "not in numbers but clearly in letters") (Smith and Karpinski). This was in large part because written numbers could be easily changed or forged; a simple flourish of a pen could transform a zero into a 6 or a 9, for instance. Roman numerals were not so easily altered; 10 is represented by the letter X, for example. Bankers recorded money orders in words, therefore, which is a practice we still utilize when writing checks today (Ghusayni 84).

Fibonacci recognized the advantages of zero and the other Arabic numerals; he knew that the benefits far outweighed the hazards. Italian merchants who agreed with him continued to use them, too, even when forbidden (Seife 81). In addition, the spread of a cashless trade society (through the issuing of bills of exchange and checks), and the growth of interest calculation compelled banks to quite pragmatically accept the new method of calculation pushed by Fibonacci. In the end, governments relented to commercial pressure and the Arabic notation flourished in Italy and soon spread throughout Europe (Seife 81).

CONTESA DI MATEMATICA: ABACISTS VS. ALGORISTS

The abacists (sometimes spelled abakists) were people who preferred to use the traditional Roman numerals and mechanical tools (abaci, boards, or checker-patterned cloths) to perform arithmetic, and algorists were people who embraced the written, symbolic Hindu-Arabic notation of place-value (including the zero) and calculated using algorithmic methods or formulas (Levy 112). Most algorists renounced the use of the abacus.

An illustration in philosopher Gregor Reisch's book, *Margarita Philosophica* (*Pearl of Wisdom*) (1503) portrays the struggle between traditional and modern methods of arithmetic. The woodcut engraving, titled "The Allegory of Arithmetic," depicts a competition of sorts between those who favored Roman numeration and clung to tradition (use of the abacus) and those who had adopted the algorithmic method and calculated on pen and paper. Banners labeled "Boetius" and "Pythagoras" identify the men in the picture. The ancient Greek scholar Pythagoras (c. 500 BCE) is shown on the right in the illustration with a worried frown, using a counting board; he represents abakists. On the left, Roman philosopher Boethius (c. 500 CE) appears to be happy as he uses Indian-Arabic numerals, representing algorists ("Fibonacci" *Famous*; "Dispute").

Classic woodcut of Arithmetica (or the Allegory of Arithmetic) supervising a contest between Boëthius, representing written calculation using Hindu-Arabic numbers, and Pythagoras, represented as using a counting board.

Of course, neither of these men were alive during Reisch's lifetime (1467 – 1525 CE); in fact, roughly 1000 years had passed between the lifetimes of the two ancient philosophers and another 1000 years had passed between Reisch and Boethius! The woodcut picture is anachronistic, belonging to a time other than that which it portrays, and the characters in the scene are symbols, representing ideas. One could say the woodcut is a kind of medieval infographic.

Between the two mathematic opponents hovers the muse of arithmetic, Arithmetica, wearing a dress adorned with the Arabic numerals. Her dress decoration and her favorable look upon the figure of Boethius suggest that, by the end of the fourteenth century, algorismi was becoming increasingly more popular (O'Shea; O'Connor and Robertson). Sometime between 1400 and 1700, it ultimately prevailed.

The adoption of the new math by European economic systems was sluggish to say the least; if it were depicted in a woodcut in Reisch's book it might be a hobbling tortoise, while the spread of the Hindu-Arabic numerals in academic circles would be a sprinting hare. Fibonacci championed the Hindu-Arabic numeral system of al-Khwarizmi and Kamil in *Liber Abaci,* which is now regarded as "the seminal work in transmitting to the West the Hindu-Arabic numerals and how to add, subtract, multiply, and divide with them." Even more influential than the encyclopedic *Liber Abaci* was his smaller, more accessible digest, the *Libro di Minor Guise* (*Book in a Smaller Manner*), which circulated widely among merchants and was copied countless times by motivated traders, merchants, and bankers (Levy 117).

Despite Fibonacci showing how useful Arabic numerals were for performing complex calculations, the printing press had not yet been invented; so, knowledge spread slowly, for the most part, during the Middle Ages. "Popes and princes and even great religious institutions possessed far fewer books than many farmers of the present age" (Smith and Karpinski). Nevertheless, as with most innovations and strategies that make profitability more efficient, the practical applications in Fibonacci's books could not help but spread like a wildfire in the tinderbox of the market economy which had developed in the Western world.

Some historians have asserted that treatises on algorism by others, such as the *Carmen de Algorismo* by Alexander de Villa Dei (c. 1240 CE) and the *Algorismus Vulgaris* by John of Halifax (Sacrobosco, c. 1250 CE) were much more influential and more widely used than Fibonacci's and "doubtless contributed more to the spread of the numerals among the common people" (Smith and Karpinski;

"*Transition*"). However, more recent research has unearthed hundreds of manuscripts called *libri d'abbaco* ("abbacus books") or *trattati d'abbaco* ("abbacus tracts") which clearly point to Leonardo's *Liber Abaci* as the "gunshot" or "spark that lit the fire of the modern commercial world" in the late Middle Ages "because it was a highly combustible landscape," already experiencing rapid commercial expansion (Devlin *Finding* 25, 27, 33).

It may now seem inconceivable that the Western world balked at adopting the new numerals embraced so "stringently" by Leonardo Pisano; they were so obviously superior to calculation methods then prevalent in Christian Europe! In twenty-first century terms, Fibonacci's *Liber Abaci* was a new market instrument of disruption because it fit an emerging market segment (international trade) that was underserved by existing tools (Roman numerals and the abacus) in the industry.

Initially, and certainly while he was alive, Fibonacci's works were intensively studied and appreciated in Italy. Copies of the practical "economic" portions of the book were handwritten and distributed by the thousands, presumably not only by merchants and traders but also by students attending the many Italian vernacular schools which suddenly appeared in the second half of the thirteenth century. Commercial mathematics (abbaco) and complex bookkeeping skills were taught in these schools, in addition to literature. Thus, *Liber Abaci* significantly influenced not only the great numbers of arithmetic tracts (trattati d'abaco) which were published after *Liber Abaci*, but also the abbaco schools which flourished in the 14th century ("Education).

BASICS OF THE FIBONACCI SEQUENCE: WHAT ARE FIBONACCI NUMBERS?

The sheer magnitude of the size of *Liber Abaci* rendered it nearly impossible (certainly impractical) to duplicate in its entirety. The English version of *Liber Abaci* (a translation begun by American Laurence Sigler and posthumously completed by his wife, Joan), "has more than 600 pages, set in a fairly small typeface; examples are what occupy most of the pages" (Devlin, *Finding* 88). The book began with explanations and illustrations of how to write and manipulate the Hindu-Arabic numbers, then Fibonacci proceeded to provide the basic mechanics of Hindu-Arabic arithmetic, which he "explain[ed] using (many) specific numerical examples, much like the way elementary school pupils are taught today (Devlin, *Finding* 116). In succeeding chapters, he supplied real-world examples and demonstrated valuable methods for solving problems specifically relevant to business and companies. The mammoth twelfth chapter contained 259 worked examples (in Sigler's translation, the chapter fills 187 printed pages) (Devlin, *Finding* 119).

Fibonacci introduced Arabic numerals in *Liber Abaci* with the simple statement: "The nine Indian figures are: 9 8 7 6 5 4 3 2 1." He then asserted that, with these nine figures, and with the sign 0 ... any number may be written (Horadam).

Leonardo reckoned that most people would have little interest in theoretical, abstract problems; they would be interested in practical applications. Therefore, Leonardo "looked for ways to dress up the abstractions in familiar, everyday clothing." He used "recreational mathematics" to introduce both Arabic numerals and the Hindu-Arabic place-valued decimal system to Europe (Devlin, *Man* 69; O'Connor and Robertson).

Since he had traveled widely and knew that "many of his fellow citizens were frequent travelers," Leonardo believed that money problems about traveling were sure to attract wide interest, so these make up his next set of examples. For his first traveler problem, he wrote:

A certain man proceeding to Luca on business to make a profit doubled his money, and he spent there 12 denari. He then left and went through Florence; he there doubled his money, and he spent 12 denari. Then he returned to Pisa, doubled his money, and spent 12 denari, and it is proposed that he had nothing left in the end. It is sought how much he had at the beginning" (Devlin, *Finding 124)*.

Other topics addressed by Leonardo in *Liber Abaci* are: multiplication and addition; subtraction; division; fractions; practical tasks and rules for trade and money; accounting; quadratic and cube roots; quadratic equations; binomials; proportion; rules of algebra; checking calculations by casting out nines; progressions; and applied algebra ("Biography").

Found on pages 123-4 of the surviving second edition of 1228 was "a theoretical family of 'abracadabric' rabbits conjured up in the mind" of the young, brilliant mathematician (Lines 6, 19). This problem leads to the introduction of the

Fibonacci numbers and the Fibonacci sequence for which Leonardo is best remembered today. He presented the following puzzle (paraphrased):

A certain man put a pair of newly-born rabbits, one male, one female, into a garden surrounded by a wall. Rabbits are able to mate at the age of one month so that at the end of its second month a female can produce another pair of rabbits. How many pairs of rabbits can be produced from that pair in a year if the rabbits never die and if every month each pair begets a new pair which from the second month on becomes productive?

He then explained: "*Because the above pair gives birth in the first month, you can double it so that after one month there are two pairs. Of these, one, that is, the first, gives birth in the second month; and so there are three pairs in the second month. Two of them become pregnant again in one month, so that in the third month two pairs of rabbits are born; and so it will be five pairs this month. Of those, three become pregnant in the same month, so there are eight pairs in the fourth month. Of these, five couples bear five pairs again; if you add them to the eight pairs, there are thirteen pairs in the fifth month.*

Of those, the five couples born this month do not mate in the same month, but the other eight couples become pregnant; and so in the sixth month there are twenty-one pairs. If you add to these the thirteen couples who are born in the seventh month, there will be thirty-four couples this month. If you add to these the twenty-one pairs born in the eighth month, there will be fifty-five pairs this month. If you add to these the thirty-four pairs born in the ninth month, there will be eighty-nine pairs this month. Add to this the fifty-five pairs born in the tenth month and this month will be 144 pairs.

Adding to these the eighty-nine pairs born in the eleventh month will be 233 pairs this month. And if you finally add to these the 144 pairs that were born last month, there are 377 pairs at the end. And so many couples will have given birth to the above-mentioned couple at the place described at the end of a year" ("The Rabbit").

We assume: 1. That a pair of rabbits has a pair of children every year. 2. These children are too young to have children of their own until two years later. 3. Rabbits never die.

Eppstein observes that the last assumption is unrealistic but makes the problem simpler: "After we have analyzed the simpler version, we could go back and add an assumption e.g. that rabbits die in ten years, but it wouldn't change the overall behavior of the problem very much."

We then express the number of pairs of rabbits as a function of time (measured as a number of years since the start of the experiment) (Eppstein):

$F(1) = 1$ -- we start with one pair

$F(2) = 1$ -- they're too young to have children the first year

$F(3) = 2$ -- in the second year, they have a pair of children

$F(4) = 3$ -- in the third year, they have another pair

$F(5) = 5$ -- we get the first set of grandchildren

The problem yields the 'Fibonacci sequence': 1, 1, 2, 3, 5, 8, 13, 21, 34, 55, . . .

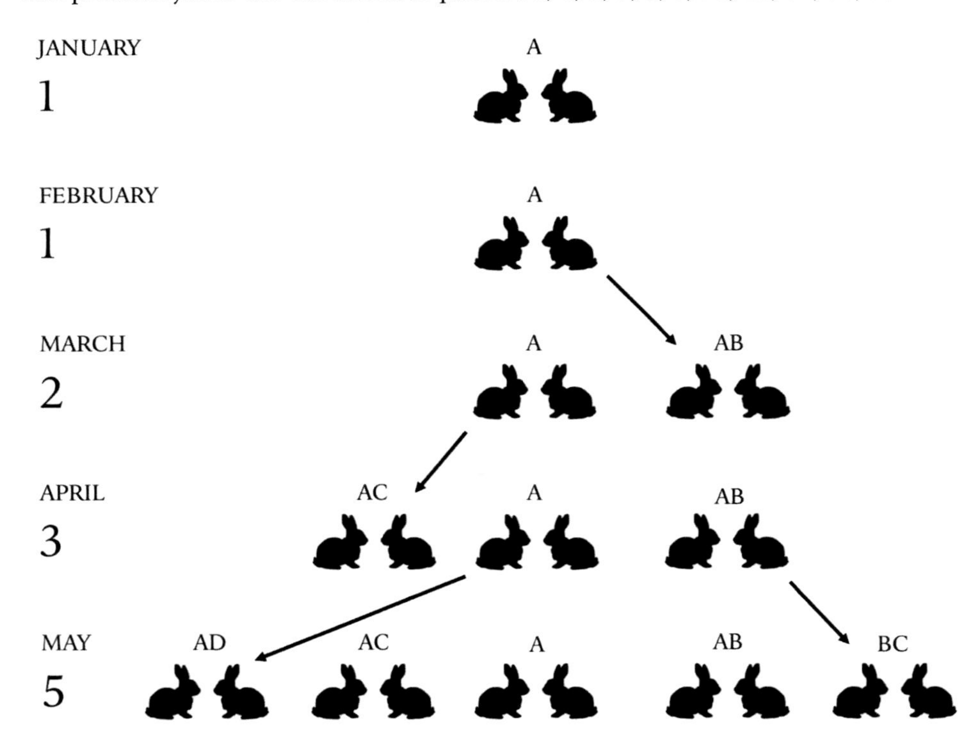

Fibonacci omitted the first term (1) in *Liber Abaci*. The recurrence formula for these numbers is: $F(0) = 0$ $F(1) = 1$ $F(n) = F(n - 1) + F(n - 2)$ $n > 1$. Although Fibonacci only gave the sequence, he obviously knew

that the nth number of his sequence was the sum of the two previous numbers (Scotta and Marketos). "This sequence, in which each number is the sum of the two preceding numbers, appears in many different areas of mathematics and science" (O'Connor and Robertson).

Fibonacci probably did not invent the rabbit problem but rather included one he had learned himself from the Moors or while traveling (Knott). He may have even relied on "translations of the works of Al-Khw^arizm^ı by Gerard of Cremona (1114-1187 CE), the latter being a pioneer in a major effort based in Toledo, Spain, to translate works written in Arabic into Latin for (Christian) Europe" (Scotta and Marketos). The sequence F(n) was already known and discussed by Indian mathematicians "who had long been interested in rhythmic patterns that are formed from one-beat and two-beat notes. The number of such rhythms having n beats altogether is F(n+1). Therefore both *Gospala* (before 1135) and *Hemachandra* (c. 1150) mentioned the numbers 1, 2, 3, 5, 8, 13, 21, ... explicitly.

"Fibonacci himself does not seem to have associated that much importance to them; the rabbit problem seemed to be a minor exercise within his work" (Scotta and Marketos). It wasn't until the 19th century that the sequence assumed "major importance and recognition thanks to the work of the French mathematician Edouard Lucas." Since then, mathematics historians have wondered about the true inspiration behind these numbers and whether Fibonacci was fully aware of their significance (Scotta and Marketos). Though Fibonacci covered a multitude of mathematical topics, he is best known for this number sequence which was later named after him by Guillaume Libri in 1838 and is still to this day being actively researched ("The Rabbit Problem").

While the "Rabbit Problem" is interesting and is the one for which he is most famous today, it is by no means the only significant mathematical problem presented in *Liber Abaci*. For example, borrowing a scenario from a ninth-century book, *Ganita Sara Sangraha*, by Mahavira (c. 800-870), Fibonacci presented a series of "purse problems" for the benefit of those who may want to divide money between two or more people. In everyday terms, he clarified the rules to follow for equal and fair distribution of something (such as money). The first solution to the "purse problem" filled half a parchment page and then he provided many more complicated variations of the same problem along with their solutions, including how to distribute the same quantity of money in a purse found by three men rather than two, a purse found by four men, and finally a purse found by five men. Pages later, Fibonacci had included solutions to eighteen different purse problems, each with a "unique twist and each using slightly different numbers." (Devlin, Finding 121).

FIBONACCI'S MATHEMATICAL CONTRIBUTIONS

First: Numbers

The place-value system of numeral writing is far easier to work with than the letter-based Roman numeral system; the position of a numeral determines its magnitude in relation to other digits in the number (for example, the 1 in 19, being the second digit in the "tens" place, signifies a value ten times its nominal value or 10 x 1). Fibonacci showed traders and merchants how to use the place-value system of arithmetic.

Second: Digits and Decimals

For centuries Europe used the Roman numeral system in which seven symbols represented seven distinct values; the Roman number 2018 could be written as MMXVIII or IIIXVMM - the letter order does not matter since the values of the letters are added to make the number.

I = 1 **V**= 5 **X** = 10 **L** = 50 **C** = 100 **D** = 500 **M** = 1000

In the Hindu-Arabic system, the order of the numerals always matters because the position of each digit determines its value; the number 2018 is quite different from 8102. Fibonacci compelled commercial use of the Arabic symbols - 1, 2, 3, 4, 5, 6, 7, 8, 9 - which had been known in Europe but had not been implemented in everyday practice; most importantly, this numeric system included a symbol for zero. Zero is needed as a place-holder because it ensures digits are placed into their proper places (columns); e.g. 2009 has no tens and no hundreds. The Roman system would have written 2009 as MMIX, omitting the values not used. Roman arithmetic was not easy; for example, MXVII added to LI is MLVIII and XLI less IV is XXXVII (Knott, "Brief").

In *Liber Abaci* (1228), Fibonacci acknowledged studying algorism extensively while traveling on business; back home in Italy, he passionately taught the rules of arithmetic he had learned from Arab mathematicians and provided the first systematic representation of the decimal system in Europe (Knott, "Brief").

Fibonacci and Algebra

While demonstrating his mathematical ability during a presentation to Emperor Frederick's Pisan court in 1225, Fibonacci explained how he would solve the following Diophantine algebraic problem: Solve $x^3 + 2x^2 + 10x = 20$. Recognizing that Euclid's method of solving equations by square roots would not work, Fibonacci used "an original method of his own, giving his answer in (Babylonian) sexagesimal notation. His approximation was far more accurate than those of his Arab contemporaries" and astounded the audience (Horadam). Soon after this occasion, Fibonacci wrote another brief treatise, *Flos* (1225) (*The Flower*, an inexplicable name since it has nothing perceivably to do with flowering plants), explaining how he had arrived at his solution to this algebra problem and another.

Before he wrote *Flos*, Fibonacci published a treatise containing solutions to algebraic equations, *Epistola ad Magistrum Theodorum*, and the book which mathematicians today consider his most important work: *Liber Quadratorum*, the *Book of Squares* (1225), a book of number theory which he dedicated to the emperor. In this book, he described the properties of the squares (such as sums of two, three or four-square numbers, or squared fractions) and tasks that lead to quadratic equations (McClenon). What makes Fibonacci's achievements even more impressive is the fact that he did not use algebraic notation as we do today because he had no such algebraic symbolism to help him. Instead, he represented numbers geometrically as line-segments, just as Euclid did. Still, his descriptions of processes and algorithms were surprisingly clear. For example, he used phrases such as res (thing) for the unknown x, and for x^2 he wrote *quadratus numerus* (square number). The following problem is representative of the type of calculation he solved and explained in a way that was superior to almost all other math textbook writers before him: "Find a square number from which, when five is added or subtracted, there always arises a square number." As Horadam states, "It is truly remarkable how far he could progress with this limited mathematical equipment. His achievements in this book justly confirm him as the greatest exponent of number theory, particularly in indeterminate analysis, in the Middle Ages" (Horadam, *Book*).

Fibonacci and Geometry

Besides the fact that rabbits produce at a "geometrical rate" (as do the numerical values of the digits in the Fibonacci sequence), there is a strange and wonderful relationship not easily seen at first glance between geometry (which deals with properties, measurement, and relationships of points, lines, angles, and figures in space) and the Fibonacci sequence, which is derived from algebra (in which symbols such as letters and numbers are combined according to the rules of arithmetic) (Scotta and Marketos).

Admittedly applying his knowledge of the theories and applications that he had learned in the "niceties of Euclid's geometric art," the medieval Italian mathematician "rediscovered" the arithmetic series (Gascueña). When a number in Fibonacci's series, (1, 1, 2, 3, 5, 8, 13, ...) is divided by the number preceding it, the quotients become the following series of numbers:

1/1 = 1, 2/1 = **2**, 3/2 = **1.5**, 5/3 = **1.666...**, 8/5 = **1.6**, 13/8 = **1.625**, 21/13 = **1.61538...**

The ratios approach the particular value called the "Golden Ratio" or the "Golden Number." It has a value of approximately **1.618034** and is represented by the Greek letter Phi (**Φ, φ**) (Scotta and Marketos).

German mathematician, astronomer, and astrologer Johannes Kepler (1571-1630 CE), (best known today for developing laws of planetary motion) noticed this pattern, in which the ratio of consecutive Fibonacci numbers approaches the Golden or Divine Ratio (Scotta and Marketos).

What is even more fascinating about this number, 1.618 ..., is the fact that it also expresses a unique relationship between two specific segments of a straight line, first defined by Euclid.

One of Euclid's strategies in *Elements* was to develop results (called propositions) about geometry which were proved solely by using logic based purely on axioms and previously-proved propositions. He showed how to divide a line in mean and extreme ratio in Book 6, Proposition 30. Euclid used this phrase to mean the ratio of the smaller part of a line (CB), to the larger part (AC): CB/AC. This ratio is the SAME as the ratio of the larger part, AC, to the whole line AB (i.e. is the same as the ratio AC/AB). Therefore, CB/AC = AC/AB. The resulting quotient is **1.618** or **Phi**, a geometric construction (concerning the properties of figures) (Gascueña).

Thus, ratios derived from the division of successive Fibonacci's numbers (from F(8) on) are the same number derived from dividing particular proportions of a straight line! (Livio 103).

While *Liber Abaci* contained some geometry problems, *Practica Geometriae* (*The Practice of Geometry*) (1223) demonstrates the mathematical brilliance of Fibonacci, for it was a well-written book containing several chapters covering basic concepts of Euclidean geometry theorems with substantial, rigorous proofs that were more advanced than the geometry of others preceding him, like Boethius and Gerbert (Pope Sylvester II). More importantly, this large book contained many practice problems dealing with area and volume formulas for plane figures and bodies (Horadam, "Eight"). The solutions and explanations, however, were not written in an esoteric language. Rather, he supplied original and instructive explanations which were widely accessible, written in the vernacular language most useful to fellow citizens; he presented solutions to problems that men in common trades (such as surveyors) were able to use to make their labor more productive and profitable. Repeatedly, Fibonacci proved to be socially relevant, which is why his achievements were clearly recognized by his contemporaries and why he is considered by modern math historians to have been a genius who was able to "see the greatness in the commonplace, and to recognize the enormous potential to change the world in what seems to most people to be a mundane or obscure idea" (Devlin, *Finding* 21).

FIBONACCI AND THE GOLDEN RATIO

Euclid's ancient ratio had been described by many names over the centuries but was first termed "the Golden Ratio" in the nineteenth century. It is not evident that Fibonacci made any connection between this ratio and the sequence of numbers that he found in the rabbit problem ("Euclid"). It was not until the late seventeenth century that the relationship between Fibonacci numbers and the

Golden Ratio was proven (and even then, not fully) by the Scottish mathematician Robert Simson (1687-1768) (Livio 101).

The Greek letter **tau (Tτ)** represented the Golden Ratio in mathematics for hundreds of years but recently (early in the 20th century) the ratio was given the symbol **phi (Φ)** by American mathematician Mark Barr, who chose the first Greek letter in the name of the great sculptor Phidias (c. 490-430 BCE) because he was believed to have used the Golden Ratio in his sculptures and in the design of the Parthenon (Donnegan; Livio 5). [The verity of these and other claims (such as that the Golden Ratio is found in paintings, Egypt's pyramids, and measurements of proportions in the human body) is addressed in "Fibonacci in Art and Music."]

German mathematician Martin Ohm (brother of physicist Georg Simon Ohm, after whom Ohm's Law is named) first used the term "Golden Section" to describe this ratio in the second edition of his book, *Die Reine Elementar-Mathematik* (*The Pure Elementary Mathematics*) (1835). He wrote: "One also customarily calls this division of an arbitrary line in two such parts the 'Golden Section.'" He did not invent the term, however, for he said, "customarily calls," indicating that the term was a commonly accepted one which he himself used (Livio 6).

The Golden Section number for phi (φ) is 0.61803 39887..., which correlates to the ratio calculated when one divides a number in the Fibonacci series by its successive number, e.g. 34/55, and is also the number obtained when dividing the extreme portion of a line to the whole. This number is the inverse of 1.61803 39887... or Phi (Φ), which is the ratio calculated when one divides a number in the Fibonacci series by the number preceding it, as when one divides 55/34, and when the whole line is divided by the largest section. The Golden Ratio formula is: F(n) = (x^n - (1-x)^n)/(x - (1-x)) where x = (1+sqrt 5)/2 ~ 1.618

Another way to write the equation is: $\varphi = \frac{1+\sqrt{5}}{2} = 1.6180339887\ldots$

Therefore, phi = 0.618 and 1/Phi. The powers of phi are the negative powers of Phi. One of the reasons why the Fibonacci sequence has fascinated people over the centuries is because of this tendency for the ratios of the numbers in the series to fall upon either phi or Phi [after F(8)]. Others have debated whether there might exist a supernatural explanation for what seems an improbable mathematical coincidence.

The limits of the squares of successive Fibonacci numbers create a spiral known as the Fibonacci spiral; it follows turns by a constant angle that is very close to the Golden Ratio. As a result, it is often called the golden spiral (Levy 121).

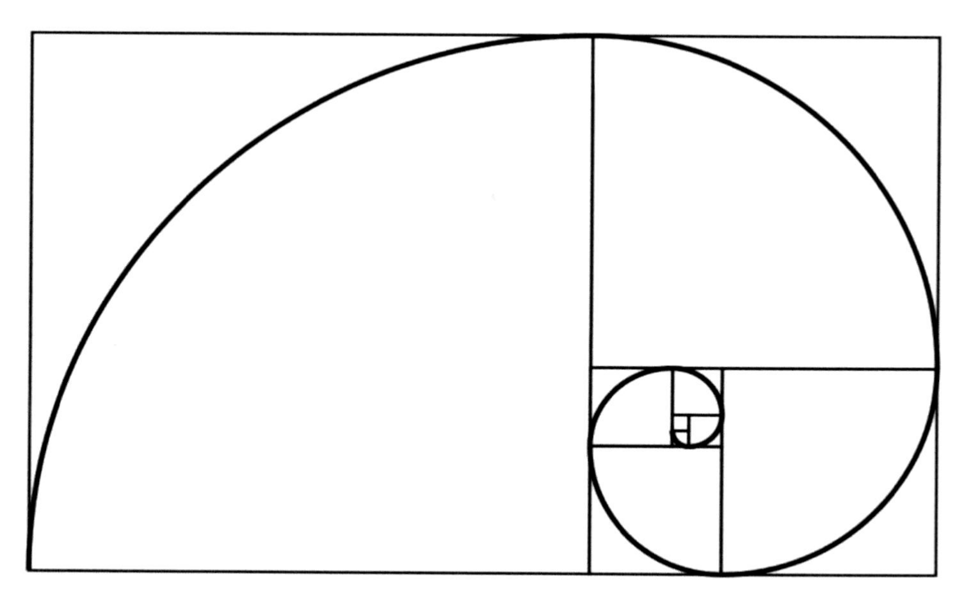

The Golden Spiral, Source: fibonacci.com

A true Golden spiral is formed by a series of identically proportioned Golden Rectangles, so it is not exactly the same as the Fibonacci spiral, but it is very similar. As the Fibonacci spiral increases in size, it approaches the angle of a Golden Spiral because the ratio of each number in the Fibonacci series to the one before it converges on Phi, 1.618, as the series progresses (Meisner, "Spirals").

Many natural phenomenon (e.g. rotations of hurricanes and the spiral arms of galaxies) and objects in nature appear to exist in the shape of golden spirals; for example, the shell of the chambered nautilus (Nautilus pompilius) and the arrangement of seeds in a sunflower head are obviously arranged in a spiral, as are the cone scales of pinecones (Knott, "*Brief*;" Livio 8).

Fibonacci spirals, Golden Spirals, and Golden Ratio-based spirals often appear in living organisms. However, not every spiral in nature is related to Fibonacci numbers or Phi; some of these spirals are equiangular spirals rather than Fibonacci or Golden Spirals. An Equiangular spiral has unique mathematical properties in which the size of the spiral increases, but the object retains its curve shape with each successive rotation. Fibonacci numbers appear most commonly in nature in the numbers and arrangements of leaves around the stems of plants, and in the positioning of leaves, sections, and seeds of flowers and other plants (Meisner, "*Spirals*").

Many observers find the patterns of Fibonacci spirals and Golden Spirals to be aesthetically pleasing, more so than other patterns. Therefore, some historians and students of math assign exceptional value to those objects and activities in nature which seem to follow Fibonacci patterns.

OTHER MATH APPLICATIONS

Fibonacci and Fractal Structures: Possibilities

Computer design specialists use algorithms to generate fractals which can produce complex visual patterns for computer-generated imagery (CGI) applications. Researchers in the Plasma Physics Research Center, Science and Research Branch, at Islamic Azad University, (Tehran, Iran) have created three variations of special fractal structures, Fibonacci fractal photonic crystals, which "could be used to develop resonant microcavities with high Q factor that can be applicable in [the] design and construction of ultrasensitive optical sensors." Possible commercial use of these structures include the production of complex visual patterns for computer-generated imagery (CGI) applications in fractal Personal Computers. Gaming enthusiasts will certainly welcome such advances in PC construction (Tayakoli and Jalili).

Fibonacci and the Physical Sciences

Kepler and others have observed Phi and Fibonacci sequence relationships between objects in the solar system and today there are websites whose curators offer propositions of their own about whether or why there are Phi relationships between the principles governing interplanetary and interstellar interactions, gravitational fields, electromagnetic fields, and many other celestial movements and forces. For example, some conclude that the Phi-related "feedback" in perturbations between the planets and the sun has the purpose of arranging the "planets into an order which minimizes work done, enhances stability and maximizes entropy" (TallBloke).

In another Fibonacci connection, neutrino physicists John Learned and William Ditto from the University of Hawaii, Mānoa, realized that frequencies driving the pulsations of a bluish-white star 16,000 light-years away (KIC 5520878) were in the pattern of the irrational "Golden Number" (Wolchover).

Atomic physicist Dr. Rajalakshmi Heyrovska has discovered through extensive research that a Phi relationship exists between the anionic to cationic radii of electrons and protons of atoms, and many other scientists have seen Phi relationships in geology, chemical structures and quasicrystalline patterns ("*Phi*;" TallBloke). The fact that such astronomically diverse systems as atoms, plants, hurricanes, and planets all share a relationship to Phi invites some to believe that there exists a special mathematical order of the universe.

GEOMETRIC CONSTRUCTIONS INVOLVING PHI

Rectangles with sides the lengths of Fibonacci numbers maintain a constant ratio (dividing the long side by the short side) no matter how large the rectangle is. Rectangles made with the Golden Ratio are called "Golden Rectangles" because many people believe them to have the most aesthetically pleasing proportions. A rectangle with sides of 8×5, for example, has a ratio of 1.6.

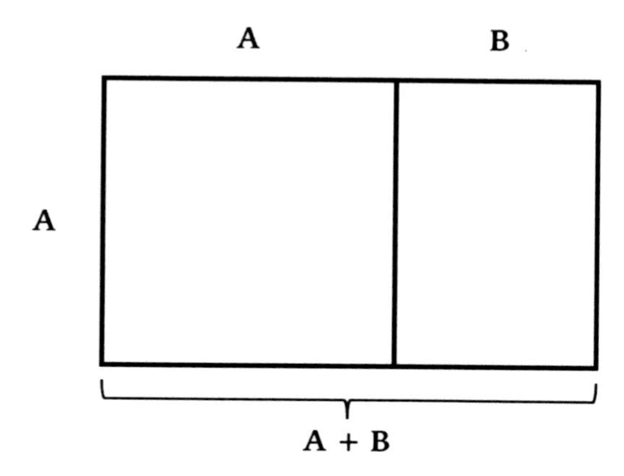

Golden Rectangle, *Source: fibonacci.com*

Golden Rectangles can be “cloned” by partitioning a square of side length equal to the length of the short side of the rectangle (Seewald), as shown previously in the image of The Golden Spiral.

Another geometric variation is the golden triangle, also known as the sublime triangle, which is an isosceles triangle in which the ratio of a side to the base is Phi.

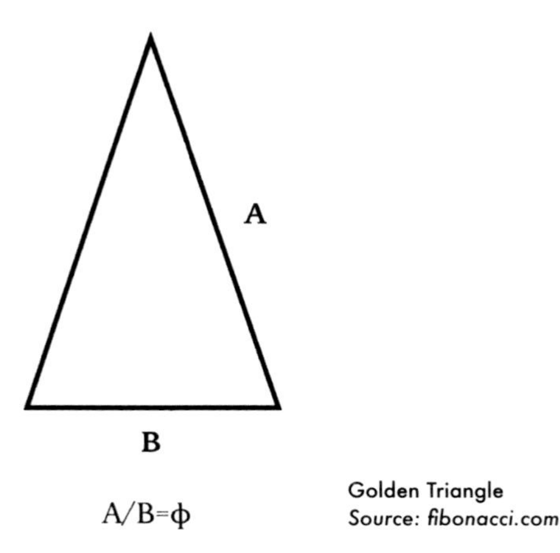

Golden Triangle
Source: fibonacci.com

In a golden triangle, a base angle of 72° can be bisected to create two additional, self-similar triangles (the internal angles and the ratios between the sides are identical no matter the length of the sides).

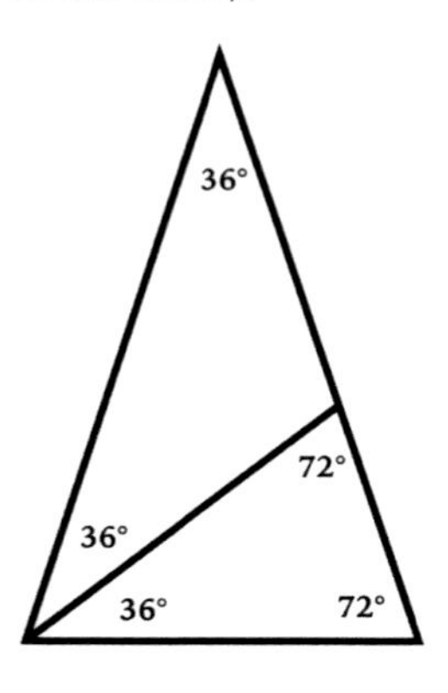

Golden Triangle & Golden Gnomon
Source: fibonacci.com

The result is an acute isosceles triangle of the same dimensions as the original, and an obtuse isosceles triangle in which the length of the equal (shorter) sides to the length of the third side creates a ratio that is the reciprocal of the Golden Ratio. This obtuse isosceles triangle is known as a "gnomon."

Pentagons, pentagrams, and decagons can be generated this way, as well (Seewald).

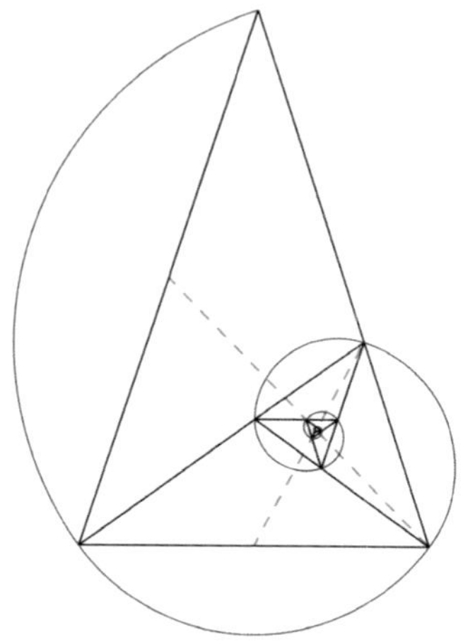

Series of Golden Triangles & Fibonacci Spiral

Scientists and mathematicians have studied these logarithmic spirals (named for the way the radius of the spiral grows when moving around it in a clockwise direction) because they recognize the patterns in objects of biology and nature, from animals and plants to vast galaxies (Seewald). They seem evident in the harmony and proportion of art and architecture, as well. The golden spiral has been utilized in the design of some modern art and architecture, but whether ancient artists and architects (such as Phidias, architect of the Parthenon) deliberately incorporated the Golden Ratio is still being debated.

FIBONACCI AND THE FUTURE

Interest in the Fibonacci family of numbers has only increased in the centuries since Leonardo Pisano was "rediscovered." Widespread interest among scientists, mathematicians, technical chartists, artists, musicians and curious intellectuals seems to geometrically intensify year after year (much like the number sequence, itself!). One reason for this popularity is the surprising frequency with which objects in nature (plants, insects, flowers) and biology repeat the patterns of the Fibonacci sequence (Lines 8).

The Fibonacci Association Foundation, based in San José, California, began in 1963 to produce the Fibonacci Quarterly, a journal devoted to the study of integers with special properties. The journal features current research in modern applications and extensions of Fibonacci's ideas. The foundation also hosts a yearly international conference (last held in July 2018 at Dalhousie University in Halifax, Nova Scotia, Canada), the purpose of which is "to bring together people from all branches of mathematics and science with interests in recurrence sequences, their applications and generalizations, and other special number sequences" (Horadam, "Eight").

PART III

III. FIBONACCI IN ART & MUSIC

ibonacci believed that calculation was an art form; to him, it was a "marvelous" thing of beauty. He considered the art of calculation with Hindu-Arabic numerals to be appealing because their use facilitates the creation of harmonious, orderly, proportionate dimensions. To a businessman like Fibonacci, order was beautiful. His proclivities were not uncommon either in his day or in ours. Modern neuroscientific research supports the ancient assumption that humans favor the aesthetic appeal of order and symmetry. Evidence suggests that "humans can detect symmetry within about 0.05 of a second. This stimulus duration is too brief for eye movements to be completed." Architect Don Ruggles, in his book titled *Beauty, Neuroscience & Architecture, Timeless Patterns and Their Impact on Our Well Being*, concludes, "this implies that human symmetry processing is a global, hard-wired activity of the brain" (Miller). The desire for harmony - one of the most ancient and primal aesthetic cravings - still exists; Fibonacci's sequence helps people objectify the subjective components of beauty ("As Easy").

Objective beauty can be more complex than bilateral symmetry or mirroring; special number sequencing and ratios are evident in diverse applications, such as literary texts (Euclid's *Elements* and Shakespearean sonnets) and architecture (the Parthenon and the Taj Mahal), botany (red rose) and sculpture (Polycleitus'

Doryphoros). The classic opinion is that beauty exists when integral parts are arranged into a coherent whole exhibiting proportion, harmony, symmetry, unity, and order. Aristotle believed the mathematical sciences uniquely demonstrate the chief forms of beauty, which are order, proportionality, symmetry, and definitude (size limitation) (*Metaphysics* vol. 2, 1705 [1078a36]) (Stakhov 40). More precisely, he proposes that "a living creature, and every whole made up of parts, must ... present a certain order in its arrangement of parts" to be considered beautiful (*Poetics*, vol. 2, 2322 [1450b34]) (Sartwell).

Aristotle's mentor, the Greek philosopher, Plato (427-347 BC), proposed a tripartite theory of soul harmony (*Republic*, c. 380 BCE), which "recognized that the highest beauty of perfect figures and proportions was based upon the principle of the division in extreme and mean ratio" (the Golden Section). This ancient Harmony theory greatly influenced the development of science and art in European culture (Stakhov 41).

For example, Roman architect and military engineer Marcus Vitruvius Pollio, better known simply as Vitruvius, wrote a treatise on the history of ancient architecture and engineering which also emphasized the importance of structural harmony (*De Architectura, On Architecture*, c. 20 BCE). Because the book is the only such work to survive intact from antiquity, it is an invaluable resource on Greek and Roman architecture, but also on a wide range of other topics such as "science, mathematics, geometry, astronomy, astrology, medicine, meteorology, philosophy, and the importance of the effects of architecture, both aesthetic and practical, on the everyday life of citizens" (Cartwright). Importantly, Vitruvius characterized architecture as embodying beauty in its complexity constrained by underlying unity. Architecture, he said, consists of order (Greek: *taxis*), arrangement (Greek: *diathesis*), proportion, symmetry, décor, and distribution (Greek: *oeconomia*, economy) (Sartwell).

Astronomer Johannes Kepler (1571-1630) expressed a similar opinion, asserting that "the chief aim of all investigations of the external world should be to discover the rational order and harmony which has been imposed on it by God and which He revealed to us in the language of mathematics" (Stakhov 42). As mathematical instruments of investigation, the Fibonacci sequence and the Golden Ratio have been used often to measure the order and harmony of some classical *oggetti d'arte e musica* (objects of art and music). Since antiquity, objects having measurable harmonious and symmetrical proportions relative to the Golden Ratio have been considered especially appealing, graceful, and beautiful.

According to Pythagoras, the most beautiful and pleasing proportion is created when a line is divided so that the ratio between the larger and smaller of the two parts is identical to that between the original line and the larger of its subdivisions, alternately called the Golden Mean or the Golden Ratio ("As Easy") or the Golden Section.

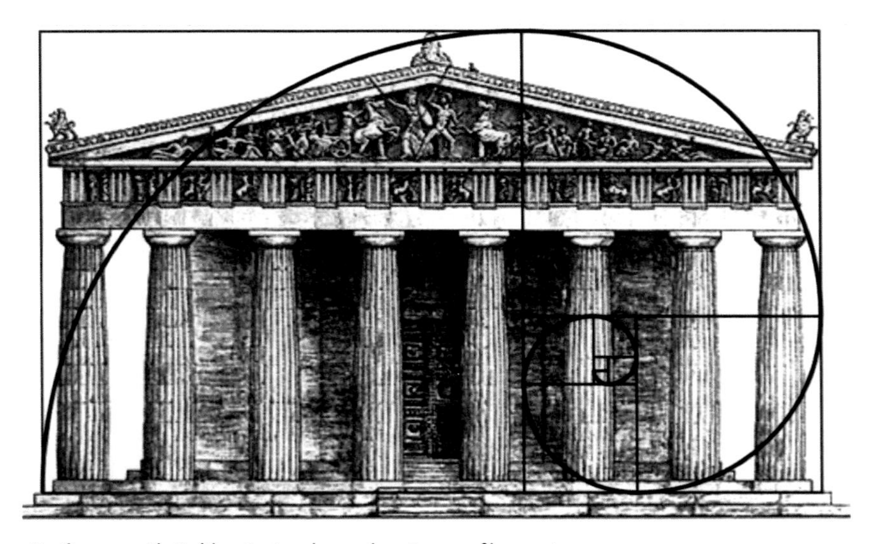

Parthenon, with Golden Rectangle overlay. *Source: fibonacci.com*

The properties of the Golden Section can be appropriated in temporal as well as spatial patterns of mathematical series and geometrical patterns (Akhtaruzzaman and Shafie). A wide survey of the relationship between Fibonacci numbers and the spheres of art and music reveals that some of the world's most outstanding artistic works incorporate design based upon the Fibonacci sequence and/or the Golden Ratio. Besides the beautiful objects mentioned previously, these include Khufu's Great Pyramid of Cheops at Giza, the sculptural Bust of Nefertiti crafted by Tuthmose, the majority of Greek sculptural monuments, the magnificent *Mona Lisa* by Leonardo da Vinci, *The School of Athens* and other works by Raphael, paintings by Shishkin and Konstantin Vasiliev, Chopin's etudes, the musical works of composers Beethoven, Tchaikovsky, Debussy and Béla Bartók, and the Modulor of Corbusier (Stakhov 59, Sinha).

FIBONACCI IN ART AND ARCHITECTURE

As previously explained, the numbers generated by Leonardo of Pisa's "rabbit problem" in Chapter 12 of *Liber Abaci* comprise a sequence that is astonishingly connected to the Golden Ratio. Ratios of successive numbers in the Fibonacci sequence (wherein each subsequent number is the sum of the previous two) become rational approximations of the Golden Mean ϕ with ever increasing accuracy. The greater the magnitude of the numbers, the more a graph of their development produces the Golden Ratio (Posamentier and Lehmann; "As Easy"). Therefore, Fibonacci numbers are used as dimensions in construction or design to "circumvent the difficulty of using the irrational number (φ)" (Posamentier and Lehmann 232).

The "modern cult of Fibonacci's numbers dates from 1877" when the French mathematician Edouard Lucas saw their significance and assigned them the name "Fibonacci sequence" ("As Easy"). Modern artists and architects are fascinated by the Fibonacci sequence, but "this is as nothing" compared to the "obsession with the Golden Ratio" in past centuries. The Golden Ratio was applied, most famously, by the architects of the Parthenon, but also by every Renaissance engineer who tried to "rediscover the lost harmony of Pythagoras" in every classical edifice emulating Phidias' magnum opus ("As Easy").

For example, evidence suggests that builders of Gothic castles used bricks formed in a special kind of rectangular parallelepiped shape which was based upon the Golden Rectangle; these were called Golden Bricks. There is speculation that the surprising strength and durability of gothic style architectural monuments is attributable to the use of the Golden Bricks (Stakhov 22). Renaissance architects, artists and designers frequently employed Golden Section proportions in eminent works of art, sculptures, paintings and architectures (Akhtaruzzaman and Shafie). According to many historians, they were nowhere near the first. In fact, many architects through the ages have either intuitively or deliberately used the Golden Section "in their sketches and construction plans, either for the entire work or for the apportionment of parts" (Posamentier and Lehmann 231).

Pythagoras, the Greek father of mathematics, is also considered to be the "father of aesthetics." Aristotle's *Metaphysics* provide proof of his mentor's beliefs concerning "a pure, clean cosmos concealed by the chaos of appearances." Pythagoras believed numbers "revealed this hidden order." It was he (or one of his followers) who first proved that "the natural notes of a plucked string only and always occur at regular intervals - when the string is subdivided at ratios of 2:1, 3:1, and so on." The same harmonies existed at "the grandest cosmic levels," Pythagoreans believed; "the music of the spheres" is a metaphysical principle derived from such Pythagorean theories ("As Easy").

Pythagoras employed arithmetic, geometric and harmonic proportions, and the law of the Golden Section. He gave exceptional consideration to the Golden Section by choosing the pentagram as the distinctive symbol of the "Pythagorean Union." Plato analyzed the five regular polyhedrons (the so-called Platonic solids) and emphasized their ideal beauty, further developing Pythagorean theories on harmony (Stakhov 40).

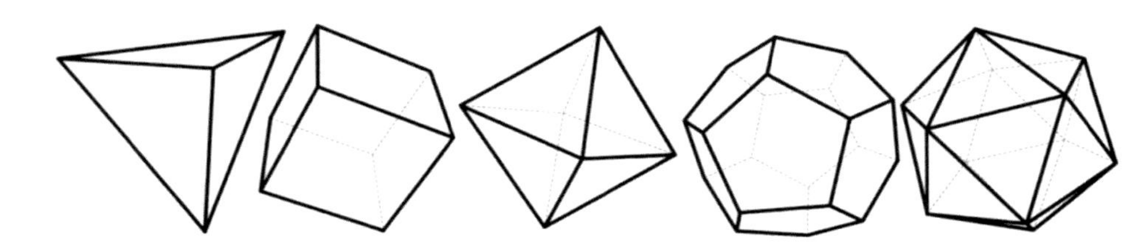

Platonic Solids, *Source: fibonacci.com*

Not long after Pythagoras, Greek mathematician Phidias (Gr. Φειδίας) (490-430 BC) appears to have applied Phi while designing the Parthenon sculptures. Two millennia before Phidias and across the Mediterranean Sea, the Egyptian engineers of the pyramids were ingenious architects who used both Pi (π) and Phi (φ) in the structural design of the Great Pyramids (Akhtaruzzaman and Shafie). The pyramids served not only as vaults of a Pharaoh's mortal remains, they were also a tribute to his majesty and power, and a monument to the riches of the country, its history and its culture. The pyramids clearly demonstrate deep "scientific knowledge" embodied in their forms, sizes, and orientation of terrain. "Each part of a pyramid and each element of its form were selected carefully to demonstrate the high level of knowledge of the pyramid creators." (Stakhov 34).

The pyramids were constructed to endure for millennia, "for all time." As the Arabian proverb says: "All in the World are afraid of a Time. However, a Time is afraid of the Pyramids" (Stakhov 34). The mathematical principle that the square of the hypotenuse of a right triangle (the side opposite the right angle) is equal to the sum of the squares of the other two sides was well-known to the Egyptians long before Pythagoras proved it (Pythagorean Theorem). They selected the golden right triangle as the geometric calibration instrument for the construction of Cheops' Pyramid. The initial height of Cheops' Pyramid is calculated to have been equal to $H = (L/2) \times \tau = 148.28$ m. The ratio of the external area of the pyramid to its base is then equal to the Golden Mean (Stakhov 36).

Fibonacci in Solar Observation

Three to five thousand years ago, around the same time the pyramids were built, Arkaim and Stonehenge were built; in the Chernihiv region, near the city of Ichnia, Ukraine, yet another manmade tool for solar observation was built using the Golden Ratio. Bezvodovka is an ancient Bronze Age architectural land monument spanning nearly twenty square kilometers. The ancient mounds of earth were thought to be burial mounds of nomadic tribes until recently, when aerial photography records and computer applications made it possible to determine their true purpose.

Oleksandr Klykavka, an agrochemist and soil scientist from the National Agricultural University in Kyiv, Ukraine, believes Bezvodovka was an ancient solar observatory like the other more famous prehistoric monuments. It is a scientific "instrument of incredible scale, the components of which are land, sky and cosmic objects," he says. At the group of mounds in the Bezvodovka plateau, the regularities of the movements of the sun and other celestial bodies are discernible on the horizon. Klykavka explains that Arkaim is located at Latitude 52°39' North, Stonehenge is located at Latitude 51°11' North, and Bezvodovka is located at Latitude 50°31' North. At a distance of nearly two thousand kilometers between them, "the three observatories are located within a single belt where the real shape of the Earth (non-ideal sphere) intercrosses with the imaginable correct shape." All three observatories have a Northern-eastern view of the sunrise on June 22 – the longest day of the year (Klykavka).

Moreover, Klykavka says, "The "distance between the center and the western distant site is specifically 830m rather than 700 or 1000m. This is significant because, if an imaginary giant Fibonacci 'Golden Spiral' could be transposed over the observatory so that "its beginning comes from the center, the spiral would proceed through a few close sites and then some distant ones." This, he says, explains why there is a distance of 2960m to some distant sites." It is a plausible explanation for why the eight original mounds (five extant) were built so distant from each other (Klykavka).

Fibonacci in Renaissance Architecture

Architectural design using the Golden Ratio or Fibonacci numbers was prevalent in Renaissance architecture. One such example is the Santa Maria del Fiore Cathedral in Florence, whose dome was constructed in 1434 by Filippo Brunelleschi (1337-1446). The rough sketch of the dome by Giovanni di Gherardo da Prato (1426) exhibits the Fibonacci numbers, 55, 89, and 144, as well as 17

S. MARIA DEL FIORE, FLORENCE
Elevation of dome and substructure. (From Sgrilli)

Source: archimaps.tumblr.com

(half the Fibonacci number 34) and 72 (half of 144) in the projection of curvature of the vault segments ("cells" or "sails") (Posamentier and Lehmann 239).

"The Apollo Project of the Golden Renaissance," by Nora Hamerman and Claudio Rossi, notes: "The pointed-fifth curve happens to be an arc of a circle whose radius is in the ratio of 8:5 to the radius of the circumscribing circle of the internal octagon base - a ratio in the Fibonacci series that closely approximates the famous Golden Section, the self-similar growth ratio . . . Similarly, the pointed-fourth curvature yields a ratio of 3/2 of the radius of the external base octagon to the radius of the vault curvature, another proportion in the Fibonacci growth series." (30)

Hidden Fibonacci numbers were recently discovered in the façade of the Church of San Nicola in southern Italy, which was already centuries old by the time the Augustinians enlarged it in the late thirteenth century, on the cusp of the Renaissance. Another eight hundred years passed before restoration efforts in 2015 made it possible to discover a series of circular and rectangular inlays in one of the church's lunettes in the portal. Professor and petrology expert at the University of Pisa, Pietro Armienti, recognized that the arrangement of the geometric figures in the marble intarsia contained a coded message while closely observing the process of marble cleaning. Professor Armienti published his research report in the "Journal of Cultural Heritage" explaining that the formerly-indecipherable artifact embodies an explicit reference to the findings of Leonardo Fibonacci ("Fibonacci Numbers").

The first nine numbers of the Fibonacci sequence, 1, 2, 3, 5, 8, 13, 21, 34 and 55, denote the radii of the various circles in the design. To Professor Armienti, the inlaid tiles "can be used as an abacus to draw sequences of regular polygons inscribed in a circle of given radius" and was "made to calculate with good approximation the sides of the regular polygons inscribed" in the largest circle ("Fibonacci Numbers").

The arabesque is inscribed within a circle, which is inscribed within a square, which is inserted in a rectangle whose ratio is the Golden Ratio. Armienti explains, "The Golden Ratio recurs in the grid of the background, now fully visible after the restoration, and it is precisely the background that provides the key to understanding the meaning of the lunette and its prominent position on the façade" (Armienti).

Armienti's article, "The Medieval Roots of Modern Scientific Thought. A Fibonacci Abacus on the Facade of the Church of San Nicola in Pisa," is a fascinating study providing many detailed descriptions of the various Fibonacci-related components of the design with explanations of how they ultimately served as a tool for "the education of elites, in tune with the aims of scholastic philosophy: a precious gift of the wisdom of the ancients whose heritage must be valued." Here are just a few of the details he provides:

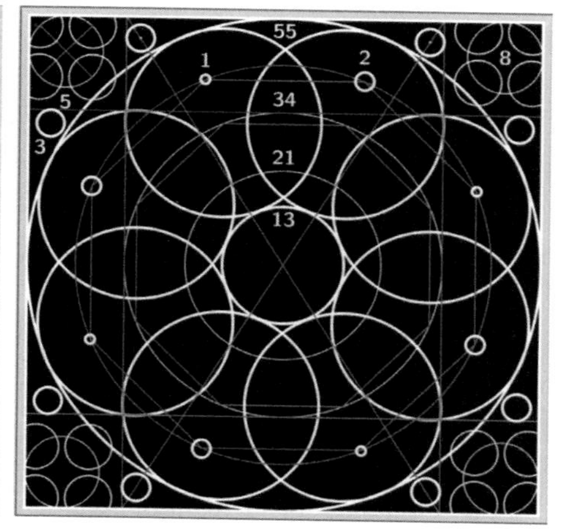

The *Lunette* of San Nicola

The concentric coronas forming a girdle in the intarsia have maximum and minimum radii of 21 and 13 respectively and have their centers on the circle of diameter 34. This arrangement describes the property that FN/FN–2 = 2 with the advance of FN-3. In fact, all the circles of size 21 in the girdle are tangent both to the circle 55 and to the circle of size 13 in the center. With reference to the diameters, this implies: 110 = 42 + 42 + 26 that is equivalent to: 55 = 21 + 21 + 13 (or 55/21 = 2 advance = 13). The same rule applies to circles of radii 34 and 13, in fact 34 = 13 + 13 + 8 (there are four coronas of size 2, between the two circles of radius 13, whose centers lie at the extremes of the radius of circle 34).

The size of the line borders in this way is necessarily 2, as required by the fact that the difference between 21 and 13 is 8 and has to be distributed on four coronas of equal size.

The other linear elements of the intarsia are inscribed in circles of 55–2 = 53 respectively (2 is the width of all the coronas), and 34–2 = 32.

Fifty-three and 32 are the sums of the N–2 Fibonacci numbers that precede 55 and 34 in the series. This arrangement is related to the second property stated above for the Fibonacci series.

Lower limits of the battlements are arranged on the circle of radius 34–2, while their cusps are defined by a kink marked by the circle of radius 42 = 55–13 along which the circles of radii 1 and 2 are also aligned.
(Armienti)

Other details, linked to the grid in the lunette around the intarsia, are proof that the author of the design was aware of the relations between the Golden Ratio and the Fibonacci series. This allowed him to find and represent to a very good approximation, regular polygons inscribed in a circle of a given radius r (Armienti).

The conclusion must be made that artists, theologians, mathematicians, and artisans worked closely together to create this masterpiece, "following a common code based on the insights of Fibonacci, and with utter dedication to their art." Armienti believes he has deciphered only a small part of this code, and there are many historiographic and mathematical problems which remain to be solved. For example, he says, "the intarsia shows that the artists were fully aware of the connections that existed between their plan, the Fibonacci sequence and the Golden Ratio, even though, until today, the discovery of these connections has been attributed to Luca Pacioli, a mathematician of the early seventeenth century" (Armienti). Clearly, Pacioli had only rediscovered what others had known – and revered - hundreds of years before.

The Golden Ratio in Sculpture

Polyclitus and Phidias are regarded as the most famous and authoritative masters of ancient Greek sculpture of the Classical era. Their statues were long considered the standard of beauty and harmonious construction. Polyclitus' statue of *Doryphorus (Spear Bearer)* (late fifth/early fourth century BCE) is considered one of the greatest achievements of classical Greek art. This statue is an archetype for the proportions of the ideal human body established by ancient Greek sculptors. The name ascribed to this sculpture is especially important because the "*Canon*" (sometimes spelled "*Kanon*") was not only a statue which deigned to display perfect human proportion, but was, in fact, a physical representation of what Polyclitus had described in his treatise on beauty, also titled "Canon." While both his written treatise and the original statue are now lost, marble copies of the statue remain and text records of observations by ancient researchers and historians enable modern vicarious examination of what some considered to be the most exquisite model of the perfect human form (Sartwell).

The ancient physician Aelius or Claudius Galenus (often Anglicized as Galen and better known as Galen of Pergamum) (130-210 CE), characterized the *Canon* as specifying the perfect symmetries of the body; he describes the statue's proportions as perfect, in "the finger to the finger, and of all the fingers to the metacarpus, and the wrist, and of all these to the forearm, and of the forearm to the arm, in fact of everything to everything... just as it is written in the *Canon* of Polyclitus" (Sartwell). Some attribute such "perfection" to the fact that Polyclitus applied the principles of the Golden Mean to his

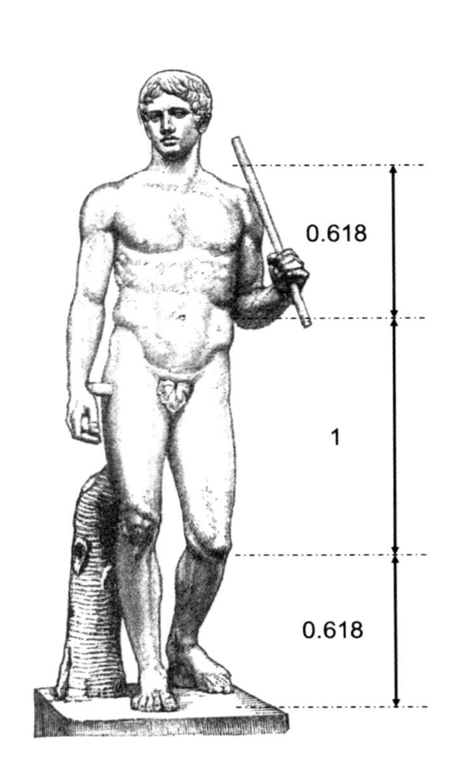

Friederichs, The Doryphoros of Polyclet (Berl. 1863); Michaelis, in the "Annali del Instituto archeologico" of 1878; *Source: fibonacci.com*

creation. Russian architect G.D. Grimm analyzed the harmonic dimensions of the *Doryphorus* and presented the following connections between the *Canon* and the Golden Mean (*Proportionality in Architecture* 1933):

1. First golden cut division: at the navel
2. Second division: lower part of the torso, passes through his knee
3. Third division: passes through the line of his neck

(Stakhov 41)

In addition to the Parthenon, Phidias created enormous statues of Athena, including one in bronze (*Winner in Battle*) and another in ivory and gold *Athena Parthenos* (*The Virgin*) in commemoration of the Athenian victory over the Persians. He also created the statue of Zeus for the Olympian temple of Zeus (about 430 B.C.), which is considered one of the Seven Wonders of the Ancient World.

Despite the unprecedented monumental sizes of his sculptures (9m *Athena Parthenos* and 13m *Zeus*), Phidias constructed them with strict adherence to the principles of harmony based on the Golden Mean (Stakhov 5-6).

Paintings, Drawings, Portraits

"Phi is more than an obscure term found in mathematics and physics." It appears to inform not only design and construction, but even our aesthetic preferences. When research subjects (who were not mathematicians or physicists familiar with phi) were asked to view random faces, those consistently deemed to be most attractive were those which exhibited "Golden Ratio proportions between the width of the face and the width of the eyes, nose, and eyebrows." Researchers conclude "the Golden Ratio elicited an instinctual reaction" (Hom).

Leonardo da Vinci's *Vitruvian Man*

Attempts to create an ideal model of a harmoniously developed human body continued during the age of the Renaissance. The ideal human figure created by Leonardo da Vinci (1452-1519) is widely known. His drawing of "Vitruvian Man" is said to illustrate the Golden Ratio (Hom). The ratio of the square side to the circle radius corresponds to the value of Phi with a deviation of just 1.7 percent (Posamentier and Lehmann 257). It clearly shows "pentagonal" or "five-fold" symmetry which is characteristic for flora and animals (Stakhov 43). The man's head, two hands, and two legs are positioned in the form of a pentagram, as if they are beams of a pentagonal star. Da Vinci based the measurements of his ideal man on the proportions described by Vitruvius (chapter 1 of Book III) (Cartwright).

Leondardo da Vinci's *Mona Lisa*, Source: fibonacci.com

Leonardo da Vinci illustrated the book *De Divina Proportione* (1509) by Franciscan monk Fra Luca Pacioli (c. 1445-1517), in which the latter referred to the number Phi as the "Divine Proportion;" Da Vinci later called this *sectio aurea*, or the Golden Section. Many assume that Da Vinci was therefore "consciously guided by this magnificent ratio" and used the Golden Ratio in all (or most) of his work (Posamentier and Lehmann 260). Some claim that he used it to define all of the proportions in his painting, *The Last Supper*, "including the dimensions of the table and the proportions of the walls and backgrounds." The Golden Ratio also appears in his iconic portrait, *Mona Lisa*. Many other famous artists are believed to have employed the Golden Ratio, including Michelangelo, (*Madonna Doni*) Raphael (*Sistine Madonna*), Rembrandt (*A Self-portrait*), Seurat (*Circus Parade*), and Salvador Dali (*Half a Giant Cup Suspended with an Inexplicable Appendage Five Meters Long*) (Posamentier and Lehmann; Hom).

It is uncommon for an artist to explicitly testify to the conscious use of Fibonacci numbers as the basic structure of their work. However, one such artist is the

German Rune Mields (b. 1935) who explained that her piece titled *Evolution: Progression and Symmetry III and IV* is "subject to the laws of symmetry" in that, in "'an ascending line, a progression of triangles is generated, with the help of the famous mathematical series of the Leonardo Pisano, called Fibonacci'" (Posamentier and Lehmann 266).

Fibonacci in Modern Architecture

Fibonacci's influence remains pervasive in Modern Architecture, where the sequence itself has become a feature of the design. The smokestack of the power station in Turku, Finland, has become a midtown landmark because it showcases the first ten numbers of the Fibonacci sequence (1, 1, 2, 3, 5, 8, 13, 21, 34, and 55) in a playful display of bright neon numbers, two meters high. The "Fibonacci Chimney" was created in 1994 by Italian artist Mario Merz as an environmental art project (Lobo). It is just one of his many "conceptual" works which incorporate the Fibonacci sequence. His "Fibonacci Naples" (1970) "consists of ten photographs of factory workers, building in Fibonacci numbers from a solitary person to a group of fifty-five." Merz featured Fibonacci's numbers because his desire was to "protest against a dehumanized, consumer-driven society" by creating art inspired by a sequence "which underlies so many growth patterns of natural life" (Livio 176).

Mole Antonelliana

Originally planned to function as a Jewish synagogue, the *Mole Antonelliana* (1863-1889) is used today as a movie museum; the five-floor building in Turin, Italy, is believed to be the tallest museum in the world. It is also Europe's tallest brick structure with the tallest dome. On one side of the four-faced dome today, the first Fibonacci numbers are illuminated by red neon lights. *Il Volo Dei Numeri* (*Flight of the Numbers*) (1998) was designed by Mario Merz ("Fibonacci – Flight").

Le Modulor

Like Fibonacci before him, 20th-century architect Charles-Edouard Jeanneret (known as Le Corbusier) (1887–1965) became fascinated by mathematical concepts while traveling; he had journeyed throughout Europe and learned principles of proportion while investigating ancient buildings everywhere he went, particularly from German architects (Cohen). Decades later, Le Corbusier published *Le Modulor: A Harmonious Measure to the Human Scale Universally Applicable to Architecture and Mechanics* and insisted his work was a unique, universally-applicable measuring system that would give architecture a mathematical order oriented to a human scale. "Le Corbusier developed his doctrine for the proportions of construction by combining the imperial measuring system based on the foot with the metric decimal system and relating this to human body measurements. He started from an assumed standard size of the human body and marked three intervals related to each other in the proportion of the Golden Ratio" ("Le Corbusier"). Specifically, he explained, "A man with a raised arm provides the main points of space displacement – the foot, solar plexus, head, and fingertip of the raised arm – three intervals which yield a number of sections that are determined by Fibonacci' (Posamentier and Lehmann 241). It was "a tool of linear or optical measures, similar to musical script," with which he was familiar (Cohen). His systematic "tool" for "planning architecture and industrial products gained worldwide currency and was applied by countless practitioners" ("Le Corbusier").

The Core

Built in the shape of a sunflower and the size of a spaceship, the Core was first built in 2005 and re-imagined in 2017-2018. Home to the Invisible Worlds exhibition at the Cornwall Education Center, the building was designed using natural forms (biomimicry) and sustainable construction and patterns based on Fibonacci numbers ("How").

Architect Jolyon Brewis explains: "We decided that the structure of the building itself should be derived from the double spiral, and we looked to the mathematics behind these spirals in nature to generate the design. We were delighted to discover that this produced an efficient and elegant network of timber beams" ("Journey" 9).

Modern Photography

San Francisco landscape photographer Mike Spinak recounts some of the many ways modern artists, including photographers, "derive a wide variety of mathematical constructs from the Golden Mean, for the sake of composition guidelines." He says:

They divide a line segment according to ~1.618 (the Golden Section). They make a rectangle where the long sides are ~1.618 times as long as the short sides (the Golden Rectangle). They make an isosceles triangle where the two long sides of the triangle are ~1.618 times the length of the short side (the Golden Triangle). They make a triangle where the longest side is ~1.618 times as long as the second longest side, which is ~1.618 times as long as the shortest side (Kepler Triangle). They make a logarithmic spiral which gets wider by a factor of ~1.618 for every quarter turn of its rotation (the Golden Spiral). And so on, with numerous others, such as the golden rhombus, Bakker's Saddle, Saint Andrew's Cross, and Bouleau's Armature of the rectangle. Some artists also like more oblique and esoteric constructs, such as division of the visible light spectrum by the Golden Mean, or segments thereof (Spinak).

Dallas photographer James Brandon offers a few suggestions for ways that amateur photographers can use the Golden Ratio to compose a photograph. According to Brandon, the software program Adobe Lightroom 3 has a Golden Ratio overlay option for cropping an image. Lines of interest or points in a photograph are lined up to coincide with a grid of the Golden Ratio. By taking the "sweet spot" of the Fibonacci Ratio and duplicating it four times into a grid, the result looks to be a rule-of-thirds grid. However, upon closer inspection it is evident that the grid does not split the frame precisely into three pieces. Instead of a 3-piece grid that divides the frame 1+1+1, there is a grid dividing the frame vertically and horizontally 1+.618+1 (Brandon).

A popular way to apply the Fibonacci spiral to a composition is to position "the primary element of the picture approximately where the tightly curled 'end' of the golden spiral would fit into the frame." The photo is "considered even more aesthetically pleasing" if the picture subject can be arranged so that "some of the picture's lines roughly follow the spiral's lines." The various other constructs listed by Spinak are occasionally "used for choosing relative proportions – such as composing with the background building ~1.618 times as tall, in the picture, as the person in the foreground. Or, they're used to choose color combinations for a picture's palette" (Spinak).

Spinak says the practice of applying the Golden Mean to composition has "seemingly become elevated to established orthodoxy." A Google Search on the topic, he adds, "will bring up more than one and a half million listings." Moreover, "most basic photography instruction books discuss composing with the Golden Mean." Adobe Lightroom has "several Golden Section overlays built into the program," including overlays for a Golden Ratio grid, a golden spiral, and a Saint Andrew's cross. Visitors to other websites can see their pictures "superimposed with Golden Mean overlays. There are also software applications available for overlaying anything on your computer screen with a variety of Golden Mean-derived visual constructs." Finally, Golden Mean calipers can be purchased online from hundreds of sources.

The Golden Mean is popular not only for composing, but also for analyzing compositions. "Analysts deconstruct pictures by drawing various line and pattern overlays upon pictures" and determining whether and how closely a particular picture conforms to some derivative of the Golden Mean (Spinak).

Brandon insists, "Fibonacci's Ratio is a powerful tool for composing your photographs, and it shouldn't be dismissed as a minor difference from the rule of thirds. While the grids look similar, using Phi can sometimes mean the difference between a photo that just clicks, and one that doesn't quite feel right." He considers Phi a far superior composition tool to use "and [a] much more intelligent and historically-proven method for composing a scene" (Brandon).

Fractals

Author Jess McNally describes fractals as "patterns formed from chaotic equations [that] contain self-similar patterns of complexity increasing with magnification." Nearly identical but smaller copies of the whole are created when you divide a fractal pattern into parts. By duplicating or repeating relatively simple fractal-generating equations, infinite complexity is formed. Unique but recognizable patterns are created. Remarkably, the number of particular geometric shapes of a specific size formed within the patterns often turn out to be Fibonacci numbers! (McNally).

FIBONACCI IN MUSIC

Musical Composition

It is not uncommon for musicians to find mathematics appealing; both disciplines involve precision, organization, and structure. Pythagoras, for example, developed musical theories based on "mathematical harmonics in frequency ratios of whole number intervals" and Galileo's father Vincenzo, a lutenist, wrote a treatise on string theory (pitch and string tension)" (Hunt). In the 17th century, Gottfried Leibniz wrote that "music is the pleasure the human mind experiences from counting without being aware that it is counting." One might say math dances with music in the mind. Indeed, Pythagoras "married" math and music when he "heard the sound of hammers on anvils and produced a formula connecting their mass to the sound they made" (Paphides). Nevertheless, while investigating possible relationships between the Golden Section and musical structure, one should not unreservedly determine the aesthetic aims of particular composers without documented testimony or material evidence, for errors are often made in measurement. The numerical value for the Golden Ratio, 5/8, is easy to confuse with the simple proportion of thirds (Fischler 31). Analysis of artwork by the cubist Juan Gris found that he may have used the diagonal of a Golden Rectangle; however, Gris categorically denied in a letter that he used the Golden atio to proportion his paintings (Fischler 31). Similar caution is warranted before drawing conclusions that any particular artist was consciously guided or inspired by the Fibonacci numbers when creating their works.

Whether or not Gris used elements of the Fibonacci sequence, diverse studies suggest composers often incorporate the golden proportion in musical compositions, perhaps due to its power in constructing well-balanced, beautiful and dynamic movements, rhythms and melodies. For example, Claude Debussy's music "contains intricate proportional systems" based on the Golden Ratio; specifically, the "dramatic climax of *Cloches a travers les feuilles* occurs when "the ratio of the total number of bars to the climax bar is approximately 1.618" (Van Gend). Fibonacci numbers harmonize naturally and the exponential growth in nature defined by the Fibonacci sequence "is made present in music by using Fibonacci notes" (Sinha). Specifically, when the Golden Section – expressed by the sequence of Fibonacci ratios – is used by a composer, it is "either used to generate rhythmic changes or to develop a melody line" (Beer 4).

"The grammar of music – rhythm and pitch – has mathematical foundations. Rhythm depends on arithmetic, harmony draws from basic numerical relationships, and the development of musical themes reflects the world of symmetry and geometry." Composers rely on symmetry to create progressions in theme and variation; they count on mathematical structure such as "prime numbers to create a sense of unease" and whispers of dissonance by creating unexpected or nontraditional rhythms and meter, such as Messiaen does in his famous *Quartet for the End of Time*. Conversely, simple ratios suggest harmony. "When we hear two notes an octave apart, the frequencies of the two notes are in an exact 1:2 ratio, so we feel we're hearing the same note." They are so similar in sound that "we give them the same name" (Du Sautoy).

Natalie Hoijer investigated compositions of concert music and compared the use of mathematical patterns such as the Fibonacci Series and the Golden Mean to other mathematical symmetries such as palindromes, crab canons, and fractals, and found that "the Fibonacci Series and Golden Mean were the most effective compositional tools and yielded the most aesthetically pleasing results" (Hoijer). This may explain why many composers, from Bartók to Debussy, have found that "the organic sense of growth found in the Fibonacci sequence of numbers is "an appealing framework" for orchestrating unique, memorable, and distinct combinations of melodic courtship (DuSautoy). Emeritus Professor of Music Theory Michael R. Rogers says, "the prevalence (and almost ubiquity) of Golden Sections throughout the so-called common-practice period is well-documented in many studies" (Rogers 249).

Beethoven

Rogers describes how Beethoven's *Piano Sonata No. 14 in C# Minor, op. 27, no. 2* serves as an archetype of golden-section form in general and represents a model of tonal clarity, the kind of temporal model with which Chopin was familiar. In the first movement of the Beethoven sonata, there are regularly expanding tonal blocks. "Each new point of arrival develops from the preceding tonal area and simultaneously prepares for the next." As each new goal cadence relates temporally both to what has gone before and to what is going to follow, "the feeling of gathering strength is inescapable." While "the arrivals are spaced further and further apart, their durational ratios to one another remain constant [and] the harmonic control is metered out" (Rogers 248).

Haylock believes he discovered a number of occurrences of Golden Ratio in the first movement of Beethoven's *Fifth Symphony*. "That's the one that starts with the famous motto theme: 'da, da, da, daah'!" He says, "In Beethoven's original score there are 600 bars before the final statement of the opening motto. But a statement of the opening motto also appears at bar 372. So, we have this structure for the three main statements of the motto: motto starts ... 372 bars ... motto starts... 228 bars ... motto starts"...divided in Golden Ratio proportion (Haylock).

Haylock mentions two other pieces of evidence he offers as proof that Beethoven used the Golden Section to create his most famous masterpiece: the 'exposition' in the first movement and the coda that is 129 bars long. "Divide this coda up using the Golden Section and you get 49 bars and 80 bars." After 49 bars of the coda, Beethoven "introduces a completely new tune that has not appeared in the movement so far. Before Beethoven no-one would ever introduce new material in a coda! So this is a very significant point in the coda." Haylock admits that we can never know whether Beethoven was "signaling his piece of radical creativity in this very long coda by linking it with the Golden Section" (Haylock).

Mozart

It is well-documented that Mozart had an early fascination with and predisposition for all things mathematical, even to the extent of loving musical gematria, or symbolic number combinations in music. "His sister Nannerl mentioned he was always playing with numbers and even scribbled mathematical equations for probabilities in the margins of some compositions (e.g. *Fantasia* and *Fugue in C Major, K394*)." Some of these may have been Fibonacci number calculations. Elements of the Golden Section appear to balance his musical lines (ratios of theme to development or musical exposition to recapitulation) and it seems likely Mozart used the Fibonacci sequence in his *Piano Sonata #1 in C major K279* as well (Hunt).

Bach

Loïc Sylvestre and Marco Costa found, through analysis of the mathematical architecture of the printed edition of Bach's compositions, that he "intentionally manipulated the bar structure of many of his collections so that they could relate to one another at different levels of their construction with simple ratios such as 1:1, 2:1, 1:2, 2:3." *The Art of Fugue* (1751), they say, shows that the whole work was conceived on the basis of the Fibonacci series and the Golden Ratio based on bar counts. For example, "distribution of Golden Ratios is evident in the numbers of bars in brackets, and in each of the subdivisions of counterpoints 8-14" (179-180).

Chopin

One of the macro-rhythmic organizational principles underpinning the harmonic and melodic ambiguities in Chopin's *Prelude in A Minor* is the Golden Section. (Rogers 245). Rogers asserts, "Golden sections are created on a melodic level within each of the first and second appearances of the minor-seventh descent" (247). Wondering why Chopin would embed one Golden Section (calculated in beats rather than measures) within another within yet another, he decides "this is a tactical timing strategy that works as a series of signals, strategically placed and deliberately paced, which regulate the harmonic ambiguities and help to foreshadow the ultimate establishment of tonal stability – a stability that arrives, in this case, just in the nick of time – at the very end." These "produce a kind of gradually emerging and increasingly focused view of tonal centricity" (248).

Handel

Examples of deliberate application of the Golden Ratio can be found in Handel's *Messiah,* which consists of 94 measures. In measure 34, after 8/13 of the first 57 measures, the entrance of the theme "The kingdom of this world..." marks an essential point (34/57). There is a very important solo trumpet entrance ("King of Kings") in measures 57 to 58, after about 8/13 of the whole piece (57/94) and another one after 8/13 of the second 37 measures, in measure 79 ("And He shall reign...") (Beer 5).

Bartók

Another of the most famous classical composers believed to have been inspired to use the Fibonacci numbers in musical composition was Hungarian Béla Bartók (1881-1945). Hungarian musicologist Erno Lendvai investigated Bartók's works "painstakingly" and published books and articles testifying that "from stylistic analyses" he was "able to conclude that the chief feature of his chromatic technique is obedience to the laws of Golden Section in every movement" (Livio 188). For example, the eighty-nine measures of *Music for Strings, Percussion and Celesta* are "divided into two parts, one with fifty-five measures and the other with thirty-four measures, by the pyramid peak" (in terms of volume) of the movement (Beers 189).

Nevertheless, other musicologists have disputed Lendvai's conclusions and it is hard to determine Bartók's true intentions when he himself "said nothing or very little about his own compositions" and "did not leave any sketches to indicate that he derived rhythms or scales numerically" (Livio 190).

Modern Music

In 2009, American Jazz artist Vijay Iyer explained why he and his trio preferred Fibonacci numbers when composing and performing music. He said the sequence's scaling property is very interesting; "because the ratios get successively closer to the Golden Ratio, the ratio 5:3 is not the same as, but 'similar' to the ratio 8:5, which is 'similar' to the ratio 13:8, or 144:89, or 6,765:4,181." His trio enjoyed exploring compositions that are "asymmetrical in a Fibonacci way: a short chord and then a long chord, three beats plus five beats, totaling eight beats." With a beat that is "standard four-four time," you could step to the beat, hearing a "chord when you take your first step, and then another chord while your knee is aloft between the second and third steps." He said, "This is a rhythm that you hear in all kinds of places – like Michael Jackson's Billie Jean" (Iyer). Working with asymmetry and "move it through Fibonacci-like transformations," the trio may "perform an asymmetric 'stretch' that maintains the same 'golden' balance over the entire measure," but they transform it a bit, trying to "preserve an 'impression' of the original – the short-and-long-ness of it – to see if [they]...can achieve that feeling of similarity" (Iyer).

Fibonacci in Instruments: The Violin

In "Stradivarius: Music of the Golden Ratio," Feng Wu voices the claim made by many that master luthier Antonio Stradivari (1644-1737) of Cremona, Italy, deliberately used the Golden Ratio to make what most people in the world consider the greatest string instruments ever created. His violins are highly prized for their tone quality and their aesthetic form (Wu). As Wu explains, Stradivari employed the system of design used by the ancient Greeks and the masters of Renaissance architecture, deriving all of his linkage dimensions and placements from a single length, the length of symmetry. "He employed the golden-section ratio and the proportion of the sides of simple triangles as well as the relationship of musical intervals to divide this length" (Bidwell). In addition, Stradivari "took special care to place the 'eyes' of the f-holes geometrically, at positions determined by the Golden Ratio" (Livio 184).

The Golden Ratio $\phi = \frac{1+\sqrt{5}}{2} = 1.61803$ can be found throughout the violin, the "Lady Blunt," by dividing lengths of specific parts of the violin (Wu):

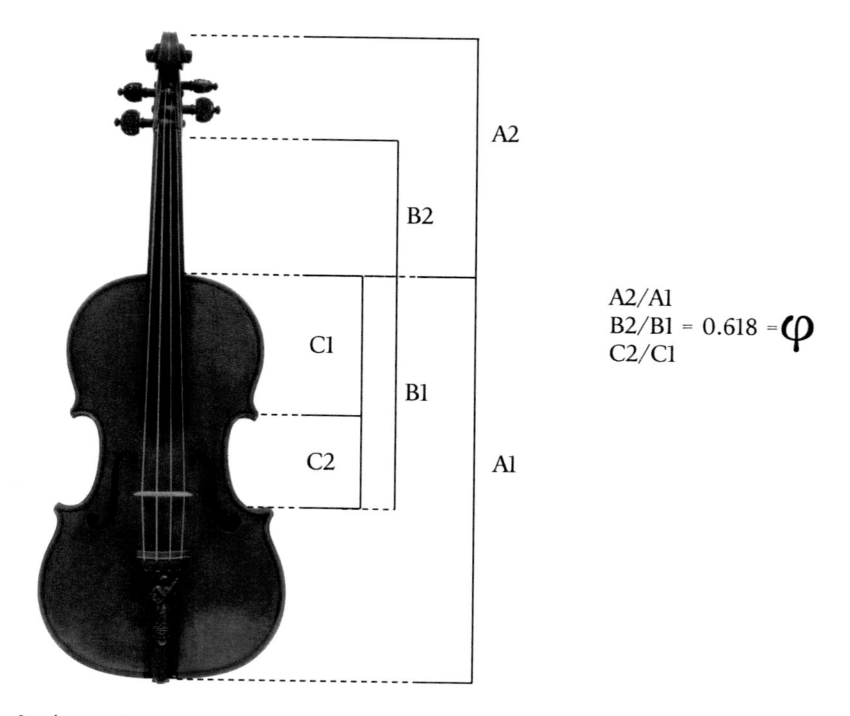

Stradivarius, "Lady Blunt" with Golden Sections Marked, *Source: fibonacci.com*

Fibonacci in Instruments: The Piano

Some claim that there is significance in the fact that the intervals between keys on a piano of the same scales are Fibonacci numbers (Sinha). The first few Fibonacci numbers appear to be represented by the arrangement of thirteen keys along the keyboard in groups of two and three black keys between the eight white which comprise a full octave (Rory). However, Livio is a member of the choir which dismisses such notion; he explains in his book, *The Golden Ratio: The Story of Phi, the World's Most Astonishing Number*, that "the chromatic scale (from C to B), which is fundamental to Western music, is really composed of twelve, not thirteen, semitones" and, more importantly, the arrangement of the keys on a piano "in two rows, with the sharp and flats being grouped in twos and threes in the upper row, dates back to the early fifteenth century, long before ... any serious understanding of Fibonacci's numbers" (Livio 185).

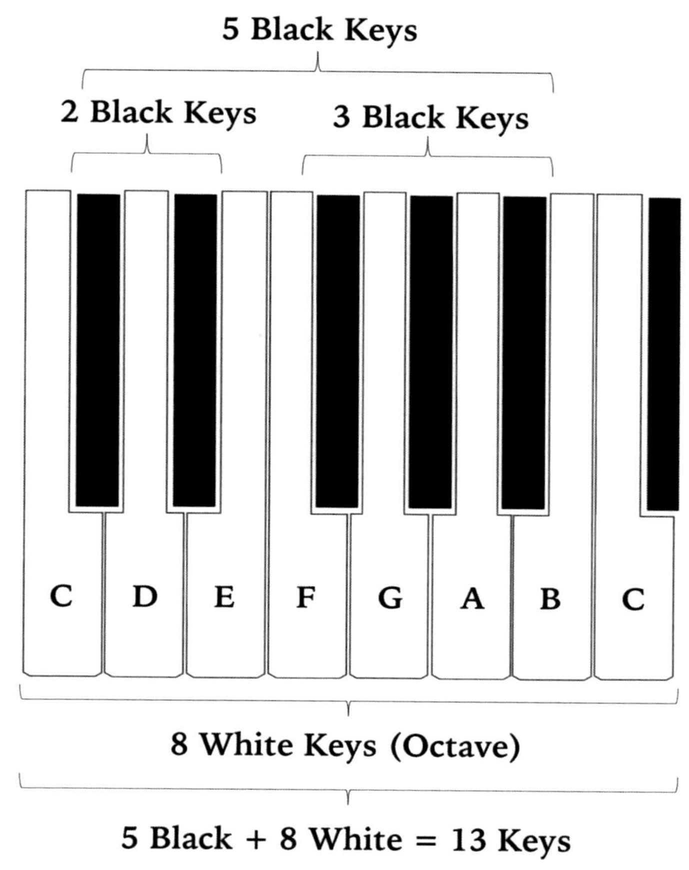

Source: fibonacci.com

FIBONACCI IN FILM

When asked by Scott MacDonald in an interview for "Film Culture" in the late seventies whether he had music in mind when he made the film *States* (1967, Revised 1970), structural filmmaker Hollis Frampton said, "It was one of the few times that I've made a real score, or graphic notation 'by the numbers." While looking for a way to order the collision of three natural substances (salt, milk, and smoke), the experimental digital artist thought about using an artificial number series because of the seeming random appearance of such collisions in nature, but then he decided to use the Fibonacci series instead. However, one problem with the Fibonacci numbers is that the series is "insufficiently dense: if you say you will put an image at frame 1, frame 2, frame 3, frame 5 and so forth, pretty quickly you have an image out around frame 1000." To avoid this problem, Frampton took not only the original Fibonacci series, but its first four harmonics. That is, he "multiplied each of the numbers by 1, 2, 3, and 4, which in musical terms would give you the fundamental, the octave, the twelfth, and the second octave." Then, he allocated three different centers – ¼, ½, and ¾ of the way through a 24,000-frame time line (1000 seconds) – for images of the substances in their gas, liquid, and solid states to spread in multiple directions. So, while Hampton said he had no particular piece of music in mind, nor even music itself, he did purposefully "use primitive procedures which are typical of music" to manipulate the harmonic series while producing his film. He explained why the Fibonacci series was useful to him in this purpose, saying, "The nice thing about the series is that it's not very symmetrical, which means that the states tended not to overlap each other" (Frampton 110).

FIBONACCI IN APPLIED ARTS

Industrial Design and Commercial Art

According to Ukrainian mathematician, inventor, and engineer Alexey Stakhov, many commercial products created today exhibit the Golden Rectangle shape in their design, including match boxes, lighters, books, credit cards, and suitcases (Stakhov 21). However, others insist some of these pop-culture beliefs are misconceptions because, for example, credit cards' "aspect ratio is defined by the ISO/IEC 7810 standard as 85.60 mm x 53.98 mm, a ratio of 1.5858:1, which differs by 2% from the Golden Mean (Salingaros, "Applications"). In addition, the television, film, and computer industries do not consistently adhere to a Golden Mean standard aspect ratio; one would think that they would "try to utilize a human preference for a specific aspect ratio." Computer screens, for example, are manufactured according to "the unofficial but ubiquitous standard of 4:3 ≈ 1.33:1." Salingaros concedes that the MacBook Pro computer's 15-inch screen does have 1,440x900 pixels, and thus an aspect ratio of 8:5 = 1.60:1. ... within 1.1% of the Golden Mean," but "the perfectionist Steve Jobs could have easily used 1,440x890 pixels for an aspect ratio of 1.6180:1 if he had wished to use the Golden Mean" (Salingaros, "Applications").

The production and arrangement of floor tiles has benefited from Phi (Φ) (1.618) which informed the development of the 1970s Penrose Tiles, allowing surfaces to be tiled in five-fold symmetry (Hom). "In the 1980s, phi (φ) (0.618) appeared in quasicrystals, a then-newly discovered form of matter whose utility for industrial purposes has only just begun to be enthusiastically investigated" (Hom).

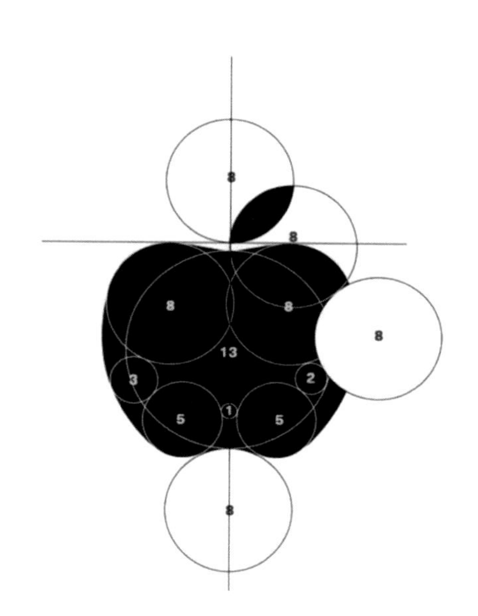

In visual arts and architecture, the Golden Section is the dynamic bearer of harmony (Stefanovic). Some have proposed that Fortune 500 companies and major companies all over the world design logos and products using the Golden Section, conceivably to satisfy consumers' inherent preference for aesthetic harmony. Apple for example, was said to have used the Golden Ratio to design its logo and many of its products (a claim which has just as often been debunked) and it's been said that Twitter used it to create their new profile page (Brandon).

Video Game Design

In 1984 two university undergraduates working out of a dormitory in Jesus College, Cambridge employed vector math to create video-game simulations of space in the spaceship game, "Elite." David Braben and Ian Bell chose not to manually plot star systems by typing coordinates of star and planets into a database; instead, Braben attempted randomly-generated numbers. However, this method led to random arrangements of the game space object representations every time the game was loaded. To overcome this problem, Braben "used the Fibonacci sequence as a seed from which identical galaxies would be generated each time the game was played, all within a computer program a fraction of the size of a photograph taken with a mobile phone today" (Parkin).

More recently, Vigil Games' senior designer Mike Birkhead explained in the 2012 article, "Tips from a Combat Designer: Fibonacci Game Design," how he uses

Fibonacci sequence numbers to limit the options for players, such as the number of weapons players are allowed to have in a game, how many talent trees, or how many monsters can be spawned in an encounter. He usually limits the choices to three or five (Fibonacci numbers). "When I think about adding something to the game I constrain myself within Fibonacci's beautiful sequence, for it forces me to REALLY commit to 'just one more,' as now it is NOT just one more, but in fact several more" (Birkhead). He finds that this self-imposed limitation upon design choices empowers him to create programs and games more confidently and efficiently.

FIBONACCI IN LITERARY ARTS

Poetry

Aristotle maintained that poetic verse expresses a beautiful impression by deliberate rhythmic, numerical relations (Stakhov 40). Just as a red rose blossoms in a well-orchestrated sequence of botanical processes governed somehow by the Golden Ratio (some say), so, too, may the literary artist deliberately conduct a poetic symphony according to elements related to the Fibonacci numbers.

Two ways the Golden Ratio and Fibonacci numbers can be used to compose poetry are: 1) There can be poems about the Golden Ratio or the Fibonacci numbers themselves or about geometrical shapes or phenomena that are closely related to them; and 2) The Golden Ratio or Fibonacci numbers can be utilized in constructing the form, pattern, or rhythm of a poem. (Livio 198)

In *Fascinating Fibonaccis,* author Trudi Hammel Garland composed a limerick comprised of five lines with the number of beats in each line being two or three, and the total number of beats being thirteen (all Fibonacci numbers):

A fly and a flea in a flue (3 beats)
Were imprisoned, so what could they do? (3 beats)
Said the fly, "Let us flee!" (2 beats)
"Let us fly!" said the flea, (2 beats)
So they fled through a flaw in the flue. (3 beats)
(Livio 198)

Fibs

Gregory K. Pincus coined the term "Fib" in 2006 to refer to a six-line, twenty syllable poem using the Fibonacci numbers and since then has maintained a popular blog, written a children's book, and has recently authored a novel, *The 14 Fibs of Gregory K.* (Pincus). The number of syllables in each line of a Fib is the sum of the previous two lines: 1, 1, 2, 3, 5, 8. "The constrained form makes you very conscious of word choice," he says (Pincus). Since 2006, Mary-Jane Grandinetti has edited "The Fib Review," an online poetry journal that specializes in only one particular poetry form – the Fibonacci poem. "Submissions are carefully selected for publication based on their poetic value and their adherence to the Fibonacci number sequence whether in syllable count, word count or any other experimental genre yet to be created" (Grandinetti).

There is no restriction for the subject of the Fibonacci poem, but the form of the Fibonacci poem is based on the structure of the Fibonacci number sequence. The poem, therefore, consists of lines with 1, 1, 2, 3, 5, 8 (and so on) syllables or words that a writer places in each line of the poem. As a literary device, it is a formatted pattern in which meaning is offered in any organized way, providing the number sequence remains the constancy of the form (Grandinetti).

"Fibonacci Salute" is a poem both inspired by and containing Fibonacci numbers in the text:

> ***Fibonacci Salute***
> *One finds*
> *one self alone suddenly where there were*
> *two when repeated percussion strikes spark*
> *three-dimensional (make that*
> *five or*
> *eight) fires with fangs in swiftest motion, scythe-like, as in an unlucky*
> *thirteen lightning strike, deal mortal blows until every*
> *twenty-one gun salute cracking the still, chill air cackles, "He'll never see*
> *thirty-four."*
>
> Shelley Allen

DISPUTATION OF FIBONACCI IN ART

Mike Spinak argues that most of the evidence presented for proof that the Golden Ratio is ubiquitous in art and nature is weak and misguided. He presents many detailed explanations for why this focus upon the use of the Golden Ratio is not beneficial for either artists, art, or patrons. He makes a case against what he says is known in academic circles as "the Golden Section hypothesis" or "the Golden Section theory."

The Golden Section theory is "a prevalent belief about composition among photographers and other visual artists" which leads to discrimination against photographers (and others) who do not follow its prescriptions. Spinak believes "the Golden Section hypothesis has been elevated to established orthodoxy," and it is unfortunate because artists and their works which do not conform to contemporary societal expectations for symmetry and compositional harmony are being discriminated against in the professional arts industry (Spinak).

"Any competent mathematician could easily derive a Golden Mean construct to match absolutely any subject placement within a frame" and interpretation is purely subjective. Unfortunately, subjective interpretation can be prejudiced (Spinak). Prejudice, of course, can be significantly detrimental to the financial health of those who are its victims, but it also harms society; for, "by accepting a flawed and rigid model as the accepted 'basis for beauty,' the biological basis for genuine beauty is replaced, and the results are unnatural" (Salingaros "Applications").

Mathematician and polymath Nikos Salingaros confirms that "the Golden Mean does indeed arise in architectural design, as a method for ensuring that a structure possesses a natural and balanced hierarchy of scales." This is the same "experimentally verified role found in a wide variety of structures in nature that exhibit hierarchical scaling" (Salingaros, "Applications"). However, researchers have also refuted claims that the Golden Mean was incorporated into design structure with paradigmatic case studies, two of which are introduced here.

The Parthenon in Athens

Salingaros joins Spinak in refuting the traditional "evidence" that the Parthenon was purposely designed or built using golden proportions. According to Salingaros, claims by artists and architects that people prefer "rectangles having aspect ratio 1.618:1 approximating the Golden Mean" are "false and are chiefly due to failing to measure things accurately." He uses perjorative terminology when he describes these as "embarrassing errors ... perpetuated by a kind of cult mysticism" (Salingaros, "Applications").

Geometrical analysis accounting for curvature of the Parthenon in 1980 determined that the design shows no evidence of use of the Golden Mean. The building was designed "based on a module of 85.76 cm" (Salingaros, "Applications").

The Parthenon's dimensions:

Eastern Width	101 ft. 3.5 in
Western Width	101 ft. 3.9 in
North Length	228 ft. 0.8 in
South Length	228 ft. 0.7 in
Height	64 ft.
Length to Width ratio	1:2.25
Height to Length ratio	1:3.56
Height to Width ratio	1:1.58

"The ancient Greeks would have built it much closer to the Golden Ratio, if they were trying to do that," Spinak asserts; after all, "they got the length on the North side of the Parthenon within about a tenth of an inch of the length on the South side" (Spinak). He also disparages the way some "Golden numberists" arbitrarily superimpose Golden Rectangles on photos of landmark buildings in order to demonstrate the presence of Golden Mean proportions.

Leonardo da Vinci

Golden "numberists" (as Spinak calls them) and adherents to the Golden Section hypothesis invariably use paintings by Leonardo Da Vinci to validate claims that Renaissance artists purposefully incorporated the Golden Mean into their works of art. However, Pacioli did not prescribe the Golden Ratio as the determinate proportion for all works of art as some allege; instead, "when dealing with design and proportion, he specifically advocates the Vitruvian system, which is based on simple (rational) ratios." It appears that French mathematicians Jean Etienne

Montucla and Jerome de Lalande were the first to falsely claim Pacioli preferred and dictated the Golden Ratio for proportion (Livio 134-5).

"*Vitruvian Man*'s navel height / full height ratio is .604, not .618" which is a difference of "a little more than two and a quarter percent." Spinak argues Leonardo would have placed the man's navel slightly higher if he had intended to maintain golden proportion. According to Spinak, another of da Vinci's iconic works is misrepresented as having been imbued with golden proportions: the *Mona Lisa*. Just as they do with the Parthenon, people place overlays of Golden Sections and Golden Rectangles atop the image and arbitrarily exclude part of her face or body to support their claims (Spinak). For example, some purport that a rectangle drawn around *Mona Lisa*'s face would have Golden Ratio dimensions; yet, Mario Livio charges, "in the absence of any clear (and documented) indication of where precisely such a rectangle should be drawn, this idea represents just another opportunity for number juggling" (162).

In conclusion, not everyone considers the Golden Mean as something to be celebrated, used, or obeyed religiously; nor are these detractors compelled to seek Fibonacci everywhere. While the idea of an amazing, mysterious, ubiquitous but purposeful pattern is appealing to some, others hold that such beliefs have no basis in reality and are "nothing more than superstition and hoax. They are not scientific observation based on evidence; they are mystical beliefs in numerology" and there is no need to embellish the magnificent splendor of nature or, for that matter, art (Spinak).

CHART OF TERMS

Term	Defined or Named By	Symbol/Value	Definition
Golden Mean	Euclid	A C B	Point C is between points A and B so that the ratio of the short part of the segment (BC) to the long part (AC) equals the ratio of the long part (AC) to the entire segment (AB)
Golden Ratio Phi Golden Proportion Divine Proportion Golden Section	Euclid Barr (re Phidias) Pacioli Da Vinci	Phi (Φ) 1.618. . . phi (φ) 0.618. . . Greek letter Tt	A straight line is in extreme and mean ratio when, as the whole line is to the greater segment, so is the greater to the lesser. 1. Ratio of smaller part of a line (CB) to larger part (AC): CB/AC = ratio of larger part (AC) to the whole line (AB). Therefore, CB/AC = AC/AB. 2. Ratio between two successive Fibonacci numbers (after F8)
Golden Angle		137.5°	1. Angle subtended by smaller arc when two arcs of a circle are in Golden Ratio 2. Angle that divides a full angle in a Golden Ratio (but measured in opposite direction so that it measures less than 180 degrees)
Golden Rectangle			1. Ratio of length to width is the Golden Ratio, 1:phi 2. Sides the lengths of Fibonacci numbers
Golden Triangle Sublime Triangle			1. Ratio of a side to base of an isosceles triangle is phi 2. Base angle 72° bisected creating two self-similar internal triangles 36°
Golden Spiral			Logarithmic spiral whose growth factor is φ (about 6.9) for every quarter turn it makes

Source: fibonacci.com

PART IV

IV. FIBONACCI IN NATURE

he Fibonacci sequence of numbers forms the best whole number approximations to the Golden Proportion, which, some say, is most aesthetically beautiful to humans. "Empirical investigations of the aesthetic properties of the Golden Section date back to the very origins of scientific psychology itself, the first studies being conducted by Fechner in the 1860s" (Green 937). Debate remains as to whether or not humans naturally prefer Golden Ratio (1.61803...) proportions in the organization and structural symmetry of art, music or nature, and some even deny that the Golden Ratio is as ubiquitous in nature as others proclaim.

Nevertheless, mathematical principles do appear to govern the development of many patterns and structures in nature, and as time passes, more and more scientific research finds evidence that the Fibonacci numbers and the Golden Ratio are prevalent in natural objects, from the microscopic structure proportions in the bodies of living beings on Earth to the relationships of gravitational forces and distances between bodies in the universe (Akhtaruzzaman and Shafie).

FIBONACCI IN BOTANY

As it relates to the development and structure of a plant, it is not uncommon to find representations of the Fibonacci numbers or the Golden Ratio. Structural symmetry is one of the simplest ways an organism will demonstrate this fascinating phenomenon (Livio 115). For example, pentagonal symmetry (five parts around a central axis, 72° apart) is quite common in the natural world, particularly among the more "primitive" phyla, such as the water net (*Hydrodictyaceae Hydrodictyon*), a green algae ("Live"). Higher in the plant kingdom, many flowers exhibit Fibonacci-number petal symmetry, including fruit blossoms, water lilies, brier-roses and all the genus *rosa*, honeysuckle, carnations, geraniums, primroses, marsh-mallows, campanula, and passionflowers. Besides symmetrical number and arrangement of parts or petals, plants often illustrate the Fibonacci sequence in their seed sections or in the spirals that are formed as new parts and branches grow.

Flowers, Fruits, and Vegetables

Spanish poet Salvador Rueda (1857-1933) eloquently said, "las flores son matematicas bellas, compass, armonia callada, ritmo mudo," (flowers are a beautiful mathematics, compass, silent harmony, mute rhythm) (Spooner 38).

Many flowers display figures adorned with numbers of petals that are in the Fibonacci sequence:

1 petal: White Calla Lily
2 petals: Euphorbia
3 petals: Lily, Iris, Euphorbia
5 petals: Buttercup, wild Rose, Larkspur, Columbine (Aquilegia), Hibiscus
8 petals: Delphiniums, Bloodroot
13 petals: Ragwort, Corn Marigold, Cineraria, Black Eyed Susan
21 petals: Aster, Shasta Daisy, Chicory
34 petals: Plantain, Pyrethrum, Daisy
55, 89 petals: Michaelmas Daisies, the Asteraceae family
(Sinha; Akhtaruzzaman and Shafie)

One of the largest families of the vascular plants, *compositae*, contains nearly 2000 genera and over 32,000 species ("Plant List") of flowering plants. *Compositae* (or *Asteraceae*) is commonly referred to as the aster, daisy, composite, or sunflower family. Family members are distributed worldwide and have a recognizable, "unique disc-shaped inflorescence, composed of numerous pentamerous florets packed on an involucrate head, surrounded by ray florets (petals) on the outside." The numbers of ray-florets and disc-florets vary from one plant to another, but they all are all "beautiful phyllotactic configurations" due to the arrangement of seeds in the seed head.

Daisy with Equiangular Spiral

The head of a composite displays definite equiangular spirals running counter-clockwise and clockwise. These bi-directional spirals intersect each other, such as: 2/3, 3/5, 5/8, 8/13, 13/21, 21/34, ... The numerators or the denominators of this series are recognizable as the Fibonacci sequence. The petal counts of Field Daisies are usually thirteen, twenty-one or thirty-four and, in the close-packed arrangement of tiny florets in the core of a daisy blossom, we can see the equiangular spiral phenomenon clearly as twenty-one counterclockwise spirals swirl in delicate, picturesque motion with thirty-four clockwise spirals. In any daisy, the floral tango of logarithmic spirals generally consists of successive terms of the Fibonacci sequence (Britton; Livio 112).

Passion Flower with 3 Sepals and 5 Outer Green Petals

The seeds are packed this way on the seed head presumably to "reduce the size of the florets to the optimum [size] necessary for quick production of an adequate number of single-seeded fruits" (Majumder and Chakravarti). The distribution of the ray-florets on the heads in Fibonacci number structure is indicative of "perfect growth," according to Majumder and Chakravarti. Research also indicates that "individual flowers emerge at a uniform speed at fixed intervals of time along a logarithmic spiral, with an initial angle at = 137.5° (Mathai and Davis, 1974)" (Majumder and Chakravarti).

Passion flower, also known as *Passiflora Incarnata,* is a perfect example of a flower regally displaying Fibonacci Numbers, for three sepals protect the bud at the outermost layer, while five outer green petals are followed by an inner layer of five more, paler, green petals. With an array of purple and white stamens, there are 5 greenish T-shaped stamens in the center and three deep brown carpels at the uppermost layer (Akhtaruzzaman and Shafie).

Sunflower with 89 Clockwise and 55 Counterclockwise Parastichies

Insufficient data and "careless methodological practices" cause many scientists to doubt or outright refute the notion that Fibonacci numbers or the Golden Ratio are an absolute "law of nature" (Green 937). Jonathan Swinton and Erinma Ochu aimed to remedy the lack of scientific evidence by studying and recording the occurrence of Fibonacci structure in the spirals (parastichies) of 657 sunflower (*Helianthus annuus*) seed heads at the MSI Turing's Sunflower Consortium. The sunflower has 55 clockwise spirals overlaid on either 34 or 89 counterclockwise spirals, a phi proportion (Phi Φ =1.618 ...) (Wright). The most reliable data subset of 768 clockwise or anticlockwise parastichy numbers revealed a clear indication of a dominance of Fibonacci structure: 565 were Fibonacci numbers and 67 had a predefined type of Fibonacci structure. They also found "more complex Fibonacci structures not previously reported in sunflowers" and seed heads without Fibonacci structure (nearly 20%). Some seed heads without Fibonacci structure nevertheless had a tendency for counts to cluster near the Fibonacci number; in those, "parastichy numbers equal to one less than a Fibonacci number were present significantly more often than those one more than a Fibonacci number." The research also revealed the "existence of quasi-regular heads, in which no parastichy number could be definitively assigned" (Swinton and Ochu).

Apple Cross Section

The bumps and hexagonal scales (bracts) on the surface of pineapples form three distinct spirals in increasing steepness, creating a recognizable pattern of Fibonacci numbers (usually 5, 8, and 13) and the Romanesco Broccoli (looks and tastes like a cross between broccoli and cauliflower) has a shape almost like a pentagon with florets organized in spirals in both directions around the center point, where the florets are smallest (Posamentier and Lehmann; Knott). Other fruits have Fibonacci numbers in their seeds' sectional arrangements. Three sections are easy to see in the cut cross-sections of the Banana, Cantaloupe, Cucumber, Kiwano fruit (African cucumber), and Watermelon. Star Fruit, Okra, and Apple seeds are arranged in a pentagram shape of five sections (Akhtaruzzaman and Shafie).

Spirals, Branches, and Leaves

According to Scotta and Marketos, the Fibonacci spiral is "fundamental to organic life." They admit that it is "not always clear why these numbers appear," but it appears that they "reflect minimization or optimization principles of some sort, namely the notion that nature is efficient yet 'lazy,' making the most of available resources" (Scotta and Marketos). New growth may simply form spirals so that the new leaves, petals, and branches will not block older leaves, etc. from sunlight or air, or so that the maximum amount of rain or dew will get directed down to the roots (Akhtaruzzaman and Shafie). Others suggest the logarithmic spiral may be a "natural outcome of the supply of genetic material in the form of pulses at constant intervals of time and obeying the law of fluid flow" (Majumder and Chakravarti).

Artichoke Flower Head Cluster

Spiral Aloe

In the world of nature, things grow by adding some unit, even if the unit is as small as a molecule. Michael Wright says phi is "an ideal rate of growth for things which grow by adding some quantity," such as the nautilus and sunflower (Wright). In addition to pineapples, the nautilus, and the sunflower, spirals are found in pinecones, ginger plants, artichokes, and other plants, including numerous cacti (Britton; Livio 110-111).

In 1868, Wilhelm Hofmeister suggested that new cells destined to develop into leaves, petals, etc. (primordia) "always form in the least crowded spot" on the meristem (growing tip of a plant). Each successive primordium of a continuously growing plant "forms at one point along the meristem and then moves radially outward at a rate proportional to the stem's growth" (Seewald). The second primordium grows as far as possible from the first, and the third grows at a distance farthest from both the first and the second primordia (Seewald). In the 1830s, scientist brothers found that the rotation tends to be an angle made with a fraction of two successive Fibonacci Numbers, such as 1/2, 1/3, 2/5, 3/8 (Akhtaruzzaman and Shafie). "As the number of primordia increases, the divergence angle eventually converges to a constant value" of 137.5° thereby creating Golden Angle Fibonacci spirals (Seewald).

The fact that branches and leaves of plants follow certain mathematical growth patterns was first noted in antiquity by Theophrastus (ca. 372 B.C. – ca. 287 B.C.) but the phenomenon was first called *phyllotaxis* ("leaf arrangement" in Greek) in 1754 by the Swiss naturalist Charles Bonnet (1720-1793) (Livio 109-110). Patterns with the other fractions are also observed, though uncommonly (Okabe). Professor Emeritus H. S. M. Coxeter at the University of Toronto, in his *Introduction to Geometry* admits that some plants exhibit phyllotaxis numbers that "do not belong to the sequence of f's [Fibonacci numbers] but to the sequence of g's [Lucas numbers] or even to the still more anomalous sequences 3,1,4,5,9,... or 5,2,7,9,16,...." He concludes we must face the fact that phyllotaxis is really not a universal law but only a fascinatingly prevalent tendency favored by nature (Coxeter).

The Sneezewort is a simple plant that exhibits the Fibonacci sequence. New shoots commonly spring from the main stem at an axil. Horizontal lines drawn through the axils highlight obvious stages of development in the plant. The pattern of development mirrors the growth of the rabbits in Fibonacci's classic problem; that is, the number of branches at any stage of development is a Fibonacci number. "Furthermore, the number of leaves in any stage will also be a Fibonacci number" (Britton).

Palms are ideal specimens of the plants that display spiral phyllotaxis because their large leaves are prominently arranged (and therefore easily observed) on the trunk. Palm leaves are arranged in Fibonacci sequence spiral formation, overlap least and provide an "angular deflection between consecutive leaves that, together, comprise a photosynthetic surface optimally accessible to illumination" (Davis; Majumder and Chakravarti).

The initial leaves are often 180° apart. As the stem matures, it thickens, and the spiral pitch between leaves decreases. "The result of this process is that angular divergence of new leaves gradually approximates the golden angle. This gives rise to an approximate logarithmic spiral of touching leaves" (Green). On the oak tree, for example, the branch rotation is a Fibonacci fraction, 2/5, which means that five branches spiral two times around the trunk to complete one pattern. Other trees with the Fibonacci leaf arrangement are the elm tree (1/2), the beech (1/3), the willow (3/8) and the almond tree (5/13) (Livio 113-115).

Okabe refers to Fibonacci phyllotaxis as evidence of natural selection, which eliminates plants whose growth patterns do not turn following the Golden Angle

$2\pi a0 = 137.5°$ He says those which follow this process are "favored in nature" because the Golden Angle is structurally "the most stable" because it undergoes the least (though inevitable) phyllotactic structural changes (stepwise transitions between phyllotactic fractions) during early stages of the growing process to a mature plant (Okabe).

FIBONACCI IN INSECTS

Nature's language is empirically mathematical. As Salvador Rueda says in "La Nitidula," the insect's "dos vuelos descurbrn/sus hermeticas palabras" (two wings disclose/its hermetic words). Simple observation of the body sections of ants and millipedes, the wing dimensions and location of eye-like spots on moths, and the beautiful design of butterfly wings reveal shapes related to the Golden Ratio (Meisner). But even more intricate biological aspects of some insects illustrate properties of the Fibonacci sequence and the Golden Ratio.

For example, from studies of the sensory reaction and attendance frequency of plant pollinators and their relationships with flowers, Leppik found that most pollinating insects have the ability to distinguish angular-form and radial-symmetry in flowers (Leppik). Even more remarkable is the fact that some insects have powers not limited to the recognition of mathematical shape and structure; they are capable of creating such harmonious structure, as well. Indeed, Rueda poetically describes the inhabitants of the beehive as geometers (skilled in geometry):

Llenas de logaritmos sus celulas obscuras,
encierran de un misterio la gran filosofia,
y rie entre sus mallas armonicaas y puras
el cuerpo rubio y ritmico que encierra la ambrosia.

(Filled with logarithms its dark cells
enclose in mystery the great philosophy,
and the golden rhythmic army that surround the ambrosia
laugh away their harmonious pure mazes.)
(Spooner)

Honey bees, like certain fish, construct their honeycombs in the form of a hexagonal lattice (Spooner 38). Their hives as a whole are elliptical in shape

(Stakhov 27). Perhaps honeybees instinctually know that "the geometrical correlations of the golden ellipse give optimal conditions for the attainability by photons ...with minimal energetic losses," as Polish scientist Jan Grzedzielski (*Energetic and Geometric Nature Code*) found when studying the golden ellipse. Before he was murdered by the Gestapo in 1941, the physician and professor observed that the golden ellipse can be used as a geometric model for the spreading of the light in optic crystals (Stakhov 27).

Fibonacci and Honeybees: Family Tree

Speaking of bees, there are between 20,000-30,000 species of bees but the one most familiar is the honeybee, which lives in a colony called a hive. Honeybees have an unusual family tree; this is due to the fact that male honeybees have only one parent. Dr. Knott explains why:

Honeybee

The queen produces the eggs in a colony of honeybees. Worker honeybees are female and produce no eggs. Drone honeybees are male; their job is to mate with the queen. The queen's unfertilized eggs produce males, so male honeybees have a mother but no father! All female honeybees are produced from fertilized eggs, when the queen has mated with a male, so they have two parents. Females most often end up as worker honeybees; only a few are fed with a special substance called royal jelly which makes them grow into queens who will leave the hive to start new colonies by taking swarms of honeybees with them to build new nests.

So female honeybees have two parents, a male and a female, whereas male honeybees have just one parent, a female.

Dr. Knott portrays the Honeybee Family Tree as showing parents above their children, so newer generations are at the bottom and older generations are higher. Such trees are valuable because they show the lineage of ancestors (predecessors, forebears, antecedents) of the creature at the bottom of the diagram. This is different from the family tree charts of the rabbit problem, which show descendants of the original pair (progeny or offspring) at the bottom.

Family Tree of a Male Drone [Honey]Bee

He had 1 parent, a female. Since his mother had two parents, he has 2 grandparents, a male and a female. Since his grandmother had two parents but his grandfather had only one, he has 3 great-grand-parents.

The numbers of ancestors in each generation of the honeybee are Fibonacci numbers:

Number of . . .	Male Honeybee	Female Honeybee
Parents	1	2
Grandparents	2	3
Great-grandparents	3	5
Great, Great-grandparents	5	8
Great, Great, Great-grandparents	8	13

Source: fibonacci.com

FIBONACCI IN ANIMALS

The ubiquity of logarithmic spirals in the animal, bird, and plant kingdoms presents a convincing case for a cosmic character of the Golden Ratio (Boeyens and Thackeray). Livio says Fibonacci numbers are "a kind of Golden Ratio in disguise," as they are found in even microscopic places, such as in the microtubules of an animal cell. These structures are "hollow cylindrical tubes of a protein polymer" which make up the cytoskeleton. This is the structure that "gives shape to the cell and also appears to act as a kind of cell 'nervous system.' Each mammalian microtubule is typically made up of thirteen columns, arranged in five right-handed and eight left-handed structures (5, 8, and 13 are all Fibonacci numbers)." Occasionally there are double microtubules with an outer envelope consisting of 21 columns, the next Fibonacci number. Some investigators argue that microtubules are more efficient "information processors" because they are structured this way rather than with other possible structures; however, Livio admits that "the apparent connection with the Fibonacci series may be coincidental" (Livio).

German psychologist Adolf Zeising (1810 –1876) studied the skeletons of animals and the branching of their veins and nerves. He observed that there are a lot of examples of the Golden Section or Divine Proportion found in animals, fishes, and

Ram's Horns

birds, in addition to insects and snails. For example, the natural design of a Peacock's feather hints at the Golden Ratio, the eye, fins and tail of a dolphin appear to fall at Golden Sections of the length of its body, and a penguin body exhibits Golden Ratio properties (Akhtaruzzaman and Shafie). Some see the Golden Spiral in the shape of the horns of both the ram and the kudu and in the curvature of elephant tusks (Boeyens and Thackeray; Masran; D'Agnese). Animal biology would seem to follow the same spontaneous growth patterns exhibited by plants. It must be noted, however, that many contemporary scholars dismiss such (and similar) claims by Golden Ratio adherents like Ghyka, the Romanian novelist, mathematician, historian, and philosopher who said, "Diagrams of proportions, however diversely arranged, can be deciphered by the same [Golden Ratio] key." Ghyka, for example, offered a "harmonic analysis" of a thoroughbred horse which showed that ratios between the length of the leg to the vertical thickness of the body are φ. These claims should be viewed with a "healthy degree of skepticism in the absence of full and replicated scientific reports," however (Green).

Nautilus Shell

FIBONACCI IN MARINE LIFE

Md. Akhtaruzzaman and Shafie mention many sea creatures which exhibit the Golden Proportion in one form or another. For example, they describe the body of the Rainbow Trout fish as having a shape on which "three Golden Rectangles together can be fitted" and "the tail fin falls in the reciprocal Golden Rectangles and square." Additionally, the individual fins also have the Golden Section properties" (Akhtaruzzaman and Shafie).

A wide variety of sea creatures also exhibit pentagonal symmetry. For example, the sea star (Astropecten Aurantiacus), the star fish (Ophiotrix capillaris) and the sand dollar (Echinarachnius parma) exhibit five-fold symmetry (Trinajstic) which Md. Akhtaruzzaman says "has a close intimacy with Golden Ratio." In addition, the growth patterns of natural shells like Conch Shell, Moon Snail Shell, and Atlantic Sundial Shell show logarithmic spiral growth patterns of Golden Section properties or golden spiral form (Akhtaruzzaman and Shafie). The chambered nautilus (Nautilus pompilius) is a specific example of one of the marine creatures whose structure represents a spontaneous logarithmic spiral growth pattern. This pelagic marine mollusk of the cephalopod family displays the self-similarity characteristic of an intrinsic response to environmental constraint, growing larger on each spiral by phi, according to Wright (Boeyens and Thackeray; Wright).

FIBONACCI IN BIRDS

The logarithmic spiral is equiangular: if you draw a straight line from the center of the spiral (pole) to any point on the spiral curve, the line always cuts the curve at precisely the same angle. Vance A. Tucker, a biologist at Duke University in Durham, North Carolina, found that falcons appear to capitalize upon this fact by keeping a slightly curved diving trajectory while hunting prey. Rather than plummeting in a straight line, the predatory birds keep "their heads straight while keeping their target in view from the most advantageous angle," naturally following "the curve of a (highly drawn-out) logarithmic spiral" in their angle of descent (Livio 120). Hawks appear to take the same (or a similar) approach when hunting their prey (Chin).

FIBONACCI IN UNIVERSE/SPACE

The Milky Way Galaxy, the Andromeda nebula, and the spiral Galaxy M81 have spiral patterns resembling the Golden Spiral (Scotta and Marketos; Meisner). Not so obvious are the Golden Ratio properties some scientists propose exist between the distribution of planets, moons, asteroids and rings in the solar system (Boeyens and Thackeray; Akhtaruzzaman and Shafie). Some astronomy hypotheses attempt to explain solar system formation by considering the mean distances of the planets from the sun. Engineers Md. Akhtaruzzaman and Amir A. Shafie theorize that, if the measurement of orbital distances of planets is started from Mercury, the first planet of the solar system, rather than Earth, the third, and the average of the mean planet orbital distances of each successive planet is taken in relation to the one before it, the value will approximate the Golden Ratio (Akhtaruzzaman and Shafie).

	Mean distance in million kilometers as per NASA	Relative mean distances where Mercury is considered at the beginning
Mercury	57.91	1
Venus	108.21	1.86859
Earth	149.6	1.3825
Mars	227.92	1.52353
Ceres (largest object in asteroid belt)	413.79	1.81552
Jupiter	778.57	1.88154
Saturn	1,433.53	1.84123
Uranus	2,872.46	2.00377
Neptune	4,495.06	1.56488
Pluto	5,869.66	1.3058

Total Relative Mean Distance	**16.18736**
The Average	**1.61874**
Golden Ratio	**1.61803**

Akhtaruzzaman and Shafie / *Source: fibonacci.com*

Galaxy Spiral

FIBONACCI IN GEOGRAPHY AND WEATHER

Fibonacci sequence numbers and relationships between them are displayed in sea wave curves and in the tributary patterns of stream and drainage patterns and in weather patterns which sometimes very closely match the Golden Spiral, such as whirlpools and hurricanes (Scotta and Marketos; Tracy). Both Hurricane Sandy and Hurricane Katrina were said to have manifested the Golden Spiral structure (Boeyens and Thackeray).

Hurricane Spiral

FIBONACCI IN HUMANS

The same phenomena of Phi that is found in nature's objects from snail shells to the spirals of galaxies is found also in the design and structure of the human body. For example, the cochlea of the ear is a Fibonacci spiral as is the spiral of the umbilical cord. The progression of the Fibonacci numbers and ratio are well suited to describing organic growth in the human body because they have the properties of self-similarity and of "gnomonic growth;" that is, only the size changes while the shape remains constant. The majority of organs in the human body maintain their overall shape and proportions as they grow (Sacco).

Simple observation confirms that Fibonacci numbers are represented by many human parts: one trunk, one head, one heart, etc. Then there are pairs: arms, legs, eyes, ears. Three is represented by the number of bones in each leg and arm and the three main parts of the hand: wrist, metacarpus and set of fingers consisting of three phalanxes, main, mean and nail. Considering each finger individually, the lengths the phalangeal bones relate to each other according to the rule of golden proportion (Akhtaruzzaman and Shafie). These dimensions allow for the flawless execution and adaptability of "transitory movements of the digits" when grasping. Specifically, radiological and anatomical studies show that "the lengths for the index, long, and ring fingers follow a newly identified, specific mathematical pattern, the Littler series, which is closely related to the Fibonacci series" (Yetkin, et al.; Hutchison and Hutchison).

Five appendages adjoin the torso: the arms, legs and a head; five appendages are on each of these: five fingers on hands and foot; and there are five openings on the face. Through these, five senses equip the body to interact with the world around it: sight, sound, touch, taste and smell. Back to the hand, five fingers are connected to five metacarpal bones forming the basis of the palm, which is connected to the wrist structure.

Continuing the count, the human arm together with fingers consists of eight parts. There are 12 pairs of ribs but some claim (without scientific evidence) that man in the past had 13 pairs of ribs. Fourteen facial bones, six middle ear bones and the throat total 21 bones. Human backbone with the skull consists of 34 bones: Eight skull bones (Crania), 24 Vertebrae, one Sacrum and one Coccyx. The base column of human body structure therefore totals 55 (21 + 34 = 55) bones (Akhtaruzzaman and Shafie). All of these numbers - 1, 2, 3, 5, 8, 13, 21, 34 and 55 - are numbers in the Fibonacci series.

Many features of the "ideal" human face are said to have ratios equal to φ; the dimension relationships between the eyes, ears, mouth and nose, for instance. The ratio of the height of the whole head to that of the head above the nose is also said to be Phi (Akhtaruzzaman and Shafie). Other examples supposedly include the ratio between the total height of the body and the distance from head to the finger tips, and "the distances from head to naval and naval to hill." Then there is the proportion between the forearm and upper arm and the one between the hand and forearm; all of these are said to follow the rule of Golden Ratio (Akhtaruzzaman and Shafie).

Dentists interested in the health of their patients study the relationships between dental aesthetics and the golden proportion. According to Dr. Stephen Marquardt, an eminent oral surgeon in California, "the height of the central incisor is in Golden Proportion to the width of the two central incisors." Dentists have used this information when addressing "a host of dental aesthetic problems." Golden Proportion grids have been developed which show Golden Rectangle relationships between the widths and heights of eight teeth of the anterior aesthetic segment, the incisors. In addition, the four front teeth, from central incisor to premolar, are in Golden Proportion to each other. In "Maxillary and Mandibular Teeth Widths" (1985), Dr. McArthur explained that "the average ratio of upper central incisor to lower central incisor is 1.62." Soon after, in 1987, Shoemaker "wrote a series of articles promoting the use of the Golden Proportion as an adjunct to cosmetic Dentistry." The article, "Le nombre d'or" (1989) by Amoric showed "many Golden Proportions in cephalometric tracings at various stages of facial growth and also included geometrical propositions."

In that same year, *The Annals of Plastic Surgery* (1989) featured the investigations of Kawakami et al. who measured for Golden Proportion balance between the eyes, nose and mouth in the facial appearance of typical Japanese individuals and compared the ratios to measurements in Caucasians. "Each ratio was then used for pre and post-operative aesthetic analysis" (Kawakami).

Similar purpose motivated dentist Yosh Jefferson in 1996 to provide Golden Proportion diagrams, cephalometric tracings, and computer-generated photographs in "Facial Beauty - Establishing a Universal Standard" to depict an ideal structure for the human head. He believed "there are possibly billions of examples of divine proportion within the human body" and "all living organisms, including humans, are genetically encoded to develop and conform to the divine proportion." The purpose of Dr. Jefferson's study was to establish a universal standard for facial beauty regardless of race, age, or sex; but his stated intent was not to empower or support discrimination. His hope was that his work would simplify the diagnoses and treatments of facial-skeletal structure misalignments in patients, thereby improving not only their dental and physical health but (as consequence) also their emotional and psychological health (Jefferson).

Elsewhere in the human body, internal organs also exhibit Golden Ratio relationships. Belgian gynecologist Jasper Verguts, at the University Hospital Leuven, says "gynecologists can instantly tell whether a uterus looks normal or not based on its relative dimensions" (Bellos). Uterine size varies in relation to age and gravidity (the number of times a woman has been pregnant). Dr. Verguts conducted ultrasound measurements of uteruses in 5,466 non-pregnant women and created a table showing the average ratio of a uterus's length to its width for different age bands (Verguts).

The data shows that the mean length/width ratio averages 1.857 at birth, decreasing to 1.452 at the age of 91 years. At the age when women are at their most fertile, between the ages of 16 and 20, the ratio of length to width is 1.6, "a very good approximation to the Golden Ratio" (Verguts, et al.; Bellos).

According to some, even the human heart beats in Golden Proportion rhythm. Doctors Gulay Yetkin, Nasir Sivri, Kenan Yalta, and Ertan Yetkin assessed the ratio of cardiac phases (diastole and systole) in 162 healthy subjects aged 20 years to 40 years after they had rested in the supine position for fifteen minutes and found that the diastolic time interval to systolic time interval ratio was 1.611 and the R–R/diastole ratio was 1.618 (Yetkin, et al.).

In addition to its activity, the human cardiovascular system is structured according to Golden Ratio design. Ashrafian and Athanasiou found that coronary arteries are distributed sequentially in a pattern that follows the Fibonacci series, resembling phyllotaxis seen in other branches in nature. Moreover, "data from 36 species has shown that the association of cardiac diameters by the sum of the diameters of all 13 branches across these species is in the order of the Golden Ratio, 1.618" (Yetkin, et al.; Ashrafian and Athanasiou). Even "diseased atherosclerotic lesions in coronary arteries follow a Fibonacci distribution" (Yetkin, et al; Gibson, et al).

On the molecular level, the nucleotide spirals of human DNA have Fibonacci proportions. Further research is needed to discover the way in which "the crystallographic structure of DNA, stress patterns in nanomaterials, the stability of atomic nuclides and the periodicity of atomic matter depend on the Golden Ratio" (Boeyens and Thackeray). Meanwhile, recent genetic research has determined that the cross-section of microscopic double helix of DNA illustrates the Phi ratio. Each spiral of the double helix traces the shape of a pentagon. The DNA molecule "measures 34 angstroms long and 21 angstroms wide for each full

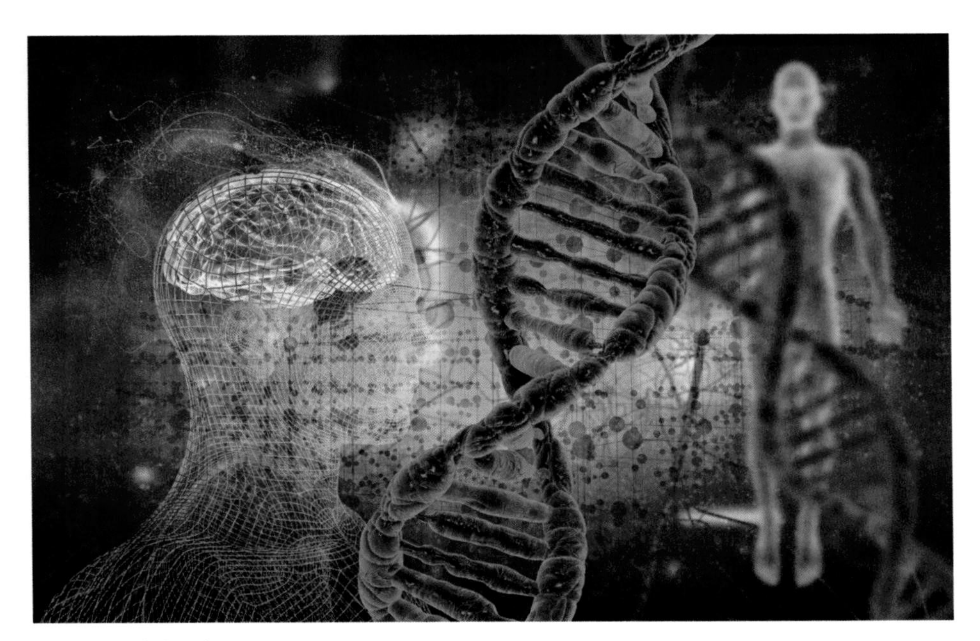

Human DNA Spiral

cycle of its double helix spiral model" so its ratio is 1.6190476, close to the ratio of Phi, 1.61803. Since the primary DNA structure molecule is formed according to Fibonacci sequence, it is assumed that linker segments between molecules are also formed according to this mathematical regularity (Shabalkin, et al.).

Just as beautiful art and music reflect harmony in nature, so, too, does the most efficient human walking pattern (gait). Roman professor and neurophysiologist Marco Iosa used a stereophotogrammetric system with 25 retroreflective markers located on the skin to analyze the spatiotemporal gait parameters of 25 healthy human subjects. Repetitive gait phases following the Golden Ratio during physiological walking were found to be most energy efficient; the ones which were "in repetitive proportions with each other," revealed "an intrinsic harmonic structure." The conclusion was that this Golden Ratio "harmony could be the key for facilitating the control of repetitive walking." Thus, harmony is not only an important component for establishing balance in art and music, it also plays a part in facilitating the maximum effective, harmonic rhythm of walking for humans (Iosa, Fusco, et al.; Iosa, Morone et al.).

FIBONACCI IN ROBOTICS AND TECHNOLOGY

The use of Fibonacci numbers to inform computer science engineering has the power to improve the lives of countless humans. Computer science and automation engineer from University of Florence (Italy) Claudio Fantacci conducted a case study involving the testing of a model of malware propagation in a computer network. He used a random generalized Fibonacci sequence to test random rates of computer infection within finite time frames because “several systems both in biology and economy are well represented by Fibonacci binary random trees” (Farina, Fantacci, and Frasca). This research is expected to help robotics engineers better anticipate and prevent disruptions in humanoid robot kinematic platforms, or robot-assisted human applications (such as the development of prostheses for loss of limb patients).

Physicist Zexian Cao and colleagues from the Chinese Academy of Sciences in China have performed stress engineering to create Fibonacci-sequence spirals on microstructures grown in the lab, and they think they have discovered the reason why the Fibonacci sequence is so ubiquitous in nature – it is a natural consequence of stress minimization (Cartwright).

They coated a curved "core" material of silver with a SiO2 shell material at a high temperature. Because material thermal expansion differs, when the composite is then cooled in a restricted geometric shape, “selective parts of the shell buckle under stress, causing patterns to form.” They created microstructures just 12 μm across and discovered that shells directed into spherical shapes during cooling developed triangular stress patterns. Forced conical shapes, however, caused spiral stress patterns to be formed. These spiral patterns had dimensions governed by the Fibonacci series, "Fibonacci spirals" (Cartwright).

This tendency may be related to something the physicist J. J. Thomson researched in 1904 when he sought “how a collection of like-charges would arrange themselves on a conducting sphere so as to minimize energy. Physicists have since calculated that the charges would take on triangular patterns - similar to Cao's spherical microstructures.” Therefore, Cao's team conclude that the Fibonacci spirals on the conical microstructures must be the equivalent minimum-energy (and hence minimum-stress) configuration for a cone. Further research and calculations need to be conducted to prove their theory (Cartwright).

Cao's experiment using pure inorganic materials may serve as the first concrete proof of the theory long held by biologists that "the branching of trees and other occurrences of the Fibonacci sequence in nature is simply a reaction to minimize stress." Creating Fibonacci patterns from stress engineering invites possible applications in photonics; Cao says, "Fibonacci spirals are a special lattice; I would say they are both ordered and disordered. If the lattice points were some materials of a proper 'dielectric,' it may provide a new photonic crystal that displays some interesting properties" (Cartwright). Photonic crystals can be used to develop biosensor technologies and materials capable of artificial touch in relation to humanoid robotics (Android) structural engineering.

CONTEMPORARY CRITICISM

While some researchers maintain that the Golden Ratio is a governing force in nature and may even be "a universal law," others conclude that the evidence presented to support the relationships between Fibonacci numbers and natural systems have been "traditionally chock through with myths, half-truths, and misconceptions" (Green). To the detractors, further scientific study is required to prove these relationships and their ubiquity. Nevertheless, some, "more carefully conducted studies have fairly consistently shown that there is, in fact, a set of phenomena that require explanation, though no one has yet produced an explanation, both adequate and plausible, that has been able to stand the test of time" (Green).

LEONARDO PISANO: HONORARIUM

To some degree, Leonardo Pisano Bigollo is a forgotten man, for he is primarily obliquely remembered by a name he did not choose. Perhaps he did not expect or desire to be known as a famous mathematician or even a teacher. After all, he deliberately referred to himself as Bigollo in some manuscripts, which could be construed to suggest he was a "man of no importance" (Venetian dialect) or, at the very least, a common traveler (Tuscan dialect) (Livio 93).

It was neither his purpose nor his goal to become famous. Perhaps this humble man methodically and meticulously compiled his mathematics instruction books only to fulfill the terms of a personal purpose which he clearly professes in his autobiographical statement in the Introduction to *Liber Abaci* (1228):

Almost everything which I have introduced I have displayed with exact proof, in order that those further seeking this knowledge, with its pre-eminent method, might be instructed, and further, in order that the Latin people might not be discovered to be without it, as they have been up to now.
(Horadam)

It is surely best to take him at his word when he testified that his intent was to provide for his extended family, his nation, and his people. By working to serve and provide for others, Leonardo's legacy endures; it is a living history.

Yes, Leonardo Pisano might be remembered primarily as the man with the "Rabbit Problem" or Fibonacci. But at the turn of the thirteenth century in Pisa, a humble mathematician tackled a significant problem stringently and provided the perfect solution: education. Through meticulous writing he equipped Europe for an evolutionary leap of economic consciousness and his instrument (knowledge) would prove to revolutionize the world.

One of the most famous professors in Italian history became an anonymous teacher, a man buried inside the consciousness of countless others, some esteemed, some mere bigollos. Eventually the "forgotten" man became the namesake for a principle in nature, of which we have not yet seen the beginning or end.

CITATIONS

PART I

"Arabic numerals in Europe." Italia Medievale. Medieval Italian Cultural Association, 3 April 2015, http://www.italiamedievale.org/portale/numerazione-araba-in-europa/?lang=en. Accessed 23 July 2018.

"Béla Bartók – The Golden Ratio in Music." ETH Zurich. ETH Library, 2018, http://www.library.ethz.ch/en/ms/Virtual-exhibitions/Fibonacci.-Un-ponte-sul-Mediterraneo/Reception-of-Fibonacci-numbers-and-the-golden-ratio/Bela-Bartok-the-golden-ratio-in-music. Accessed 20 July 2018.

"Biography of Leonardo of Pisa." ETH Zurich. ETH Library, 2018, http://www.library.ethz.ch/ms/Virtuelle-Ausstellungen/Fibonacci.-Un-ponte-sul-Mediterraneo/Biographie-von-Leonardo-Pisano. Accessed 20 July 2018.

"Calculating with Fingers." ETH Zurich. ETH Library, 2018, http://www.library.ethz.ch/en/ms/Virtual-exhibitions/Fibonacci.-Un-ponte-sul- Mediterraneo/Fibonacci-s-significance-for-the-present-day/Calculating-with-fingers. Accessed 20 July 2018.

Devlin, K. *Finding Fibonacci: The Quest to Rediscover the Forgotten Mathematical Genius Who Changed the World.* Princeton University Press, 2017.

Devlin, K. *The Man of Numbers: Fibonacci's Arithmetic Revolution.* Walker Publishing Company, New York, 2011.

The Editors of Encyclopaedia Britannica. "Pisa." Britannica.com. Encyclopædia Britannica, Inc., 2018, https://www.britannica.com/place/Pisa-Italy, Accessed 27 July 2018.

"Education in Europe - Medieval Education." Science Encyclopedia – Jrank. Net Industries, 2018, http://science.jrank.org/pages/9077/Education-in-Europe-Medieval-Education.html. Accessed 20 July 2018.

"Fibonacci." Famous Scientists. famousscientists.org, 17 July 2018, www.famousscientists.org/fibonacci-leonardo-of-pisa/. Accessed 20 July 2018.

Fitzpatrick, Richard, "Euclid's Elements of Geometry." utexas.edu. The University of Texas at Austin, 2008, http://farside.ph.utexas.edu/Books/Euclid/Elements.pdf, Accessed 29 July 2018.

Gies, F. C. "Fibonacci Biography & Facts." Encyclopaedia Britannica. Encyclopædia Britannica, Inc.,11 June 2018, https://www.britannica.com/biography/Fibonacci. Accessed 20 July 2018.

Horadam, A. F. "Eight Hundred Years Young." Vol 31, pp 123-134. The Australian Mathematics Teacher, 1975, Evansville.edu. 2018, http://faculty.evansville.edu/ck6/bstud/fibo.html. Accessed 19 July 2018.

Knott, Ron. "A Brief Biographical Sketch of Fibonacci, His Life, Times and Mathematical Achievements." Surrey.ac.uk. University of Surrey, 25 July 2005, http://personal.ee.surrey.ac.uk/Personal/R.Knott/Fibonacci/fibBio.html. Accessed 25 July 2018.

Livio, M. *The Golden Ratio: The Story of Phi, the World's Most Astonishing Number*. Broadway Books, New York, 2002.

McClenon, R. "Leonardo of Pisa and His Liber Quadratorum." The American Mathematical Monthly, Official Journal of the Mathematical Association of America. Volume XXVI. Jan. 1919, https://archive.org/details/jstor-2974039. Accessed July 2018.

O'Connor, J. J. and E. F. Robertson. "Leonardo Pisano Fibonacci." MacTutor History of Mathematics Archive, School of Mathematics and Statistics University of St Andrews, Scotland, 1998, http://www-history.mcs.st-andrews.ac.uk/Biographies/Fibonacci.html. Accessed 20 July 2018.

O'Shea, Thomas. "Typus Arithmeticae." Dr. Thomas O'Shea. 2018. https://www.drthomasoshea.com/typus-arithmeticae.html. Accessed 21 July 2018.

"The Rabbit Problem." ETH Zurich. ETH Library, 2018, http://www.library.ethz.ch/en/ms/Virtuelle-Ausstellungen/Fibonacci.-Un-ponte-sul-Mediterraneo/Bedeutung-Fibonaccis-fuer-die-Gegenwart/Kaninchenaufgabe. Accessed 20 July 2018.

Radford, L. "Reviews for Mathematics Education across Time and Place." Dr. Thomas O'Shea. 2018, https://www.drthomasoshea.com/reviews.html, Accessed 21 July 2018.

Roselaar, S. "Economic Networks and Cultural Change in the Antique Mediterranean." The Blue Web, 2018, http://bluenetworks2014.weebly.com/antiquity.html. Accessed 20 July 2018.

Scotta, T. C. and P. Marketos. "On the Origin of the Fibonacci Sequence." History. MCS. 23 Mar. 2014, http://www-history.mcs.st-and.ac.uk/~history/Publications/fibonacci.pdf. Accessed 19 July 2018.

Sesiano, Jacques (1997-07-31). "Abū Kāmil". Encyclopaedia of the history of science, technology, and medicine in non-western cultures. Springer. pp. 4–5.

Smith, David Eugene and Louis Charles Karpinski. "The Project Gutenberg EBook of the Hindu-Arabic Numerals." Gutenberg. Project Gutenberg, 14 Sept. 2007, https://www.gutenberg.org/files/22599/22599-h/22599-h.htm. Accessed 20 July 2018.

"The Transition from Roman to Hindu-Arabic Numerals." The Blue Web. 6 June 2014, http://bluenetworks2014.weebly.com/the-transition-from-roman-to-hindu-arabic-numerals.html. Accessed 20 July 2018.

PART II

"Calculating." ETH Zurich. ETH Library, 2018, http://www.library.ethz.ch/en/ms/Virtual-exhibitions/Fibonacci.-Un-ponte-sul-Mediterraneo/Fibonacci-s-significance-for-the-present-day/Calculating-with-fingers. Accessed 08 Aug 2018.

"Development of Numeral Systems." ETH Zurich. ETH Library, 2018, http://www.library.ethz.ch/en/ms/Virtual-exhibitions/Fibonacci.-Un-ponte-sul-Mediterraneo/Fibonacci-s-significance-for-the-present-day/Development-of-numeral-systems. Accessed 20 July 2018.

Devlin, K. *Finding Fibonacci: The Quest to Rediscover the Forgotten Mathematical Genius Who Changed the World.* Princeton University Press, 2017.

Devlin, K. *The Man of Numbers: Fibonacci's Arithmetic Revolution*. Walker Publishing Company, New York, 2011.

"Dispute Between Abacists and Algorists." ETH Zurich. ETH Library, 2018, http://www.library.ethz.ch/ms/Virtuelle-Ausstellungen/Fibonacci.-Un-ponte-sul-Mediterraneo/Bedeutung-Fibonaccis-fuer-die-Gegenwart/Streit-zwischen-Abakisten-und-Algoristen. Accessed 10 Aug 2018.

Donnegan, James, M.D. "A New Greek and English Lexicon: Principally on the Plan of the Greek and German Lexicon of Schneider, the Words Alphabetically Arranged." Hilliard, Gray, Boston, 1834, https://archive.org/details/newgreekenglishl00donnuoft. Accessed 23 July 2018.

"Education in Europe - Medieval Education." Science Encyclopedia – Jrank. Net Industries, 2018, http://science.jrank.org/pages/9077/Education-in-Europe-Medieval-Education.html. Accessed 20 July 2018.

Eppstein. "ICS 161: Design and Analysis of Algorithms Lecture notes for January 9, 1996." geeksforgeeks. UCI Donald Bren School of Information & Computer Sciences. https://www.ics.uci.edu/~eppstein/161/960109.html. Accessed 09 Aug 2018.

"Euclid." ETH Zurich. ETH Library, 2018, http://www.library.ethz.ch/en/ms/Virtual-exhibitions/Fibonacci.-Un-ponte-sul-Mediterraneo/Reception-of-Fibonacci-numbers-and-the-golden-ratio/Euclid-construction-of-the-golden-ratio. Accessed 9 Aug 2018.

"Fibonacci." Famous Scientists. famousscientists.org, 10 Sept 2018, www.famousscientists.org/fibonacci-leonardo-of-pisa/. Accessed 20 July 2018.

Gascueña, Dory. "Fibonacci and the Golden Ratio: Divine Geometry?" Open Mind. BBVA Group, 15 March 2017, https://www.bbvaopenmind.com/en/fibonacci-and-the-golden-ratio-divine-geometry/. Accessed 06 August 2018.

Ghusayni, Badih. "Favorite mathematics topics from the 12th Century to the 21st Century." Academic Journal Academic Journal. International Journal of Mathematics & Computer Science. Vol. 13 Issue 1, p83-104. 22p. 2018, http://content.ebscohost.com/ContentServer.asp?
EbscoContent=dGJyMMTo50SeprE4v%2BvlOLCmr1Cep7JSsqe4S7KWxWXS&ContentCust
omer=dGJyMPGqskm1r7ZJuePfgeyx9Yvf5ucA&T=P&P=AN&S=R&D=a9h&K=127054609. Accessed 3 August 2018.

Horadam, A.F. "Book Review: Leonardo Pisano (Fibonacci)—The Book of Squares (an annotated translation into modern English)—L.E. Sigler, Academic Press 1987." University of New England, Armidale, Australia 2351, p 382. https://www.fq.math.ca/Scanned/26-4/review.pdf. Accessed 13 Aug 2018.

Horadam, A. F. "Eight Hundred Years Young." Vol 31, pp 123-134. The Australian Mathematics Teacher, 1975, Evansville.edu. 2018, http://faculty.evansville.edu/ck6/bstud/fibo.html. Accessed 19 July 2018.

Knott, Ron. "A Brief Biographical Sketch of Fibonacci, His Life, Times and Mathematical Achievements." Surrey.ac.uk. University of Surrey, 25 July 2005, http://personal.ee.surrey.ac.uk/Personal/R.Knott/Fibonacci/fibBio.html. Accessed 25 July 2018.

Knott, Ron. "Dudeney's Cows." Surrey.ac.uk. University of Surrey, Guildford, Surrey GU2 7XH, United Kingdom, 25 September 2016, http://www.maths.surrey.ac.uk/hosted-sites/R.Knott/Fibonacci/fibnat. Accessed 08 August 2018.

Levy, Joel. *A Curious History of Mathematics: The (BIG) Ideas from Early Number Concepts to Chaos Theory*. Andre Deutsch, London, 2013.

Lines, Malcolm E. *Think of a Number*. Adam Hilger, New York, 1990.

Livio, M. *The Golden Ratio: The Story of Phi, the World's Most Astonishing Number*. Broadway Books, New York, 2002.

McClenon, R. “Leonardo of Pisa and His Liber Quadratorum.” The American Mathematical Monthly. Official Journal of the Mathematical Association of America, Volume XXVI. Jan. 1919, https://archive.org/details/jstor-2974039. Accessed 2` July 2018.

Meisner, Gary. “Spirals and the Golden Ratio.” GoldenNumber.net. PhiPoint Solutions, LLC, 25 Aug 2012, https://www.goldennumber.net/spirals/. Accessed 13 Aug 2018.

O’Connor, J. J. and E. F. Robertson. “Leonardo Pisano Fibonacci.” MacTutor History of Mathematics Archive, School of Mathematics and Statistics University of St Andrews, Scotland, 1998, http://www-history.mcs.st-andrews.ac.uk/Biographies/Fibonacci.html. Accessed 20 July 2018.

O’Shea, Thomas. “Typus Arithmeticae.” Dr. Thomas O’Shea. 2018. https://www.drthomasoshea.com/typus-arithmeticae.html. Accessed 21 July 2018.

“Phi in Atomic Structure.” Sacred Geometry. Drupal, 2018, http://www.sacredgeometry.es/?q=en/content/phi-atomic-structure, Accessed 10 Aug 2018.

“The Rabbit Problem.” ETH Zurich. ETH Library, 2018, http://www.library.ethz.ch/en/ms/Virtuelle-Ausstellungen/Fibonacci.-Un-ponte-sul-Mediterraneo/Bedeutung-Fibonaccis-fuer-die-Gegenwart/Kaninchenaufgabe. Accessed 20 July 2018.

Radford, L. “Reviews for Mathematics Education across Time and Place.” Dr. Thomas O’Shea. 2018, https://www.drthomasoshea.com/reviews.html, Accessed 21 July 2018.

Scotta, T. C. and P. Marketos. “On the Origin of the Fibonacci Sequence.” History. MCS. 23 Mar. 2014, http://www-history.mcs.st-and.ac.uk/~history/Publications/fibonacci.pdf. Accessed 19 July 2018.

Seewald, Nick. “The Myth of the Golden Ratio.” Golden Ratio Myth. Weebly, 2010, https://goldenratiomyth.weebly.com/a-brief-history-of-phi.html. Accessed 6 Aug 2018.

Seife, Charles. *Zero: The Biography of a Dangerous Idea*. Viking, 2000.

Sesiano, Jacques (1997-07-31). "Abū Kāmil." Encyclopaedia of the History of Science, Technology, and Medicine in Non-Western Cultures. Springer. pp. 4–5.

Smith and Karpinski. “The Hindu-Arabic Numerals, Boston and London, 1911.” Archibald, Euclid's Book on Divisions of Figures; with a Restoration based on Woepcke's Text and on the Practica Geometrice, of Leonardo Pisano. Cambridge, England, 1915.

TallBloke. “A Remarkable Discovery: All Solar System Periods Fit the Fibonacci Series and the Golden Ratio. Why Phi?” TallBloke’s Talkshop, 20 Feb 2013, https://tallbloke.wordpress.com/2013/02/20/a-remarkable-discovery-all-solar-system-periods-fit-the-fibonacci-series-and-the-golden-ratio-why-phi. Accessed 07 Aug 2018.

“The Transition from Roman to Hindu-Arabic Numerals” The Blue Web. Bluenetworks2014, 6 June 2014, http://bluenetworks2014.weebly.com/the-transition-from-roman-to-hindu-arabic-numerals.html. Accessed 10 Aug 2018.

Velasquez, Robert. “5 Reasons Using The Fibonacci Sequence Makes You Better At Agile Development.” eLearningIndustry. 7 Nov 2017, https://elearningindustry.com/using-the-fibonacci-sequence-makes-better-agile-development-5-reasons. Accessed 13 Aug 2018.

Watkins, John J. *Number Theory: A Historical Approach*. Princeton University Press, 2013.

Wolchover, Natalie. “Strange Stars Pulse to the Golden Mean.” Quanta Magazine. Quanta, 10 Mar 2015, https://www.quantamagazine.org/variable-stars-have-strange-nonchaotic-attractors-20150310/. Accessed10 Aug 2018.

PART III

Akhtaruzzaman, Md. and Amir A. Shafie. "Geometrical Substantiation of Phi, the Golden Ratio and the Baroque of Nature, Architecture, Design and Engineering," International Journal of Arts, Vol. 1 No. 1, 2011, pp. 1-22. doi: 10.5923/j.arts.20110101.01.

Armienti, Pietro. “The Medieval Roots of Modern Scientific Thought. A Fibonacci Abacus in the Facade of the Church of San Nicola in Pisa Ietro.” 2015, Journal of Cultural Heritage (2015), http://dx.doi.org/10.1016/j.culher.2015.07.015. Accessed 20 Aug 2018.

“As Easy as 1, 1, 2, 3 ....” The Guardian. Guardian News and Media Limited, 12 May 2005, https://www.theguardian.com/music/2005/may/12/classicalmusicandopera1. Accessed 22 Aug 2018.

Beer, M., 2008. “Mathematics and Music: Relating Science to Arts?” Mathematical Spectrum, 41(1):36-42. file:///C:/Users/jrllightfilter/Downloads/mathandmusic.pdf. Accessed 20 Aug 2018.

Bidwell, Eugene Boyd. “Violin Design Method of Stradivari.” Violincad.com. 2018, http://www.violincad.com/violin/betts/article.htm. Accessed 05 Sep 2018.

Birkhead, Mike. “Tips from a Combat Designer: Fibonacci Game Design.” Gamasutra.com. UBM, 15 Feb 2012, https://www.gamasutra.com/view/news/129642/Tips_from_a_combat_designer_Fibonacci_game_design.php. Accessed 21 Aug 2018.

Brandon, James. “Divine Composition with Fibonacci’s Ratio (The Rule of Thirds on Steroids).” Digital Photography School. Digital Photography School, 2018, https://digital-photography-school.com/divine-composition-with-fibonaccis-ratio-the-rule-of-thirds-on-steroids/. Accessed 19 Aug 2018.

Cartwright, Mark. “Vitruvius.” Ancient History Encyclopedia. Ancient History Encyclopedia Limited, 22 April 2015, https://www.ancient.eu/Vitruvius/. Accessed 20 Aug 2018.

Cohen, J. -L. (2014). *Le Corbusier’s Modulor and the Debate on Proportion in France*. Architectural Histories, 2(1), Art. 23. DOI: http://doi.org/10.5334/ah.by, https://journal.eahn.org/articles/10.5334/ah.by/. Accessed 22 Aug 2018.

Du Sautoy, Marcus. “Listen by Numbers: Music and Maths.” The Guardian. Guardian News, 27 Jun 2011, https://www.theguardian.com/music/2011/jun/27/music-mathematics-fibonacci. Accessed 18 Aug 2018.

Edmark, John. “John Edmark/Work/Golden Angle.” John Edmark. John Edmark, 2018, http://www.johnedmark.com/work/golden-angle/. Accessed 22 Aug 2018.

“The Fibonacci Numbers Have Been Discovered on a Church in Pisa.” 18 Sep 2015, https://sandrozicari.com/2015/09/18/the-fibonacci-numbers-have-been-discovered-on-a-church-in-pisa/. Accessed 20 Aug 2018.

“Fibonacci – The Flight of the Numbers.” Arte.it. ART.it, 23 Aug 2018. http://www.arte.it/work_of_art/fibonacci-the-flight-of-the-numbers-5739. Accessed 23 Aug 2018.

Fischler, Roger. “On the Application of the Golden Ratio in the Visual Arts.” Leonardo, vol. 14, no. 1, 1981, pp. 31–32. JSTOR, JSTOR, www.jstor.org/stable/1574475. Accessed 19 Aug 2018.

Frampton, Hollis, and Scott MacDonald. "Interview with Hollis Frampton: The Early Years." October, vol. 12, 1980, pp. 103–126. JSTOR, JSTOR, www.jstor.org/stable/778577. Accessed 20 Aug. 2018.

Grandinetti, Mary-Jane. "The Fib Review." Muse-Pie Press. 30 Jun 2018, http://www.musepiepress.com/fibreview/index.html. Accessed 21 Aug 2018.

Haylock, Derek. "Golden Ratio and Beethoven's 5th." Derek-Haylock.blogspot.com http://derek-haylock.blogspot.com/2014/08/golden-ratio-and-beethovens-5th.html. Accessed 20 Aug 2018.

Hoijer, Natalie. "Unleashing Music's Hidden Blueprint: An Analysis of Mathematical Symmetries Used in Music." IWU.edu, Digital Commons @ Illinois Wesleyan University. 2015, https://digitalcommons.iwu.edu/cgi/viewcontent.cgi?article=1007&context=music_papers. Accessed 20 Aug 2018.

Hom, Elaine J. "What Is the Golden Ratio?" LiveScience.com. Purch, 24 Jun 2013, https://www.livescience.com/37704-phi-golden-ratio.html. Accessed 21 Aug 2018.

Horadam, A. F. "Eight Hundred Years Young." Vol 31, pp 123-134. The Australian Mathematics Teacher, 1975, Evansville.edu. 2018, http://faculty.evansville.edu/ck6/bstud/fibo.html. Accessed 19 July 2018.

"How We Built the Core." Eden Project.com. Eden Project, https://www.edenproject.com/eden-story/behind-the-scenes/how-we-built-the-core. Accessed 20 Aug 2018.

Hunt, Patrick. "Mozart and Mathematics." Electrum Magazine. Electrum Magazine, 30 June 2013, http://www.electrummagazine.com/2013/06/mozart-and-mathematics/. Accessed 20 Aug 2018.

Iyer, Vijay. "Strength in Numbers: How Fibonacci Taught Us How to Swing." The Guardian. Guardian News and Media Limited, 15 Oct 2009, https://www.theguardian.com/music/2009/oct/15/fibonacci-golden-ratio. Accessed 22 Aug 2018.

"Journey to the Core." Eden Project.com. Eden Project, 2018, https://www.edenproject.com/sites/default/files/documents/journey-to-the-core.pdf. Accessed 20 Aug 2018.

Klykavka, Oleksandr. "Solar Observatory in Ukraine." Ancient History Et Cetera. Ancient History et cetera, 15 Dec 2016, http://etc.ancient.eu/education/solar-observatory-ukraine/. Accessed 21 Aug 2018.

"Le Corbusier – the Modulor." ETH Library. ETH Library, 2018, http://www.library.ethz.ch/en/ms/Virtual-exhibitions/Fibonacci.-Un-ponte-sul-Mediterraneo/Reception-of-Fibonacci-numbers-and-the-golden-ratio/Le-Corbusier-the-Modulor. Accessed 22 Aug 2018.

Livio, Mario. *The Golden Ratio: The Story of PHI, the World's Most Astonishing Number*. Broadway Books; Reprint edition, 2003.

Lobo, Carmen. "Fibonacci Met Reinfried Marass, but Who is Fibonacci?" ART & Thoughts. ART & Thoughts, 10 Feb 2013, https://articulosparapensar.wordpress.com/2013/02/10/fibonacci-met-reinfried-marass-but-who-is-fibonacci/. Accessed 22 Aug 2018.

McNally, Jess. "Earth's Most Stunning Natural Fractal Patterns." Wired.com. Conde Nast, 9 Oct 2010, https://www.wired.com/2010/09/fractal-patterns-in-nature/. Accessed 22 Aug 2018.

Meisner, Gary. "Acoustics and the Golden Ratio." Golden Number.net. PhiPoint Solutions, LLC, 2018, https://www.goldennumber.net/acoustics/. Accessed 21 Aug 2018.

Miller, Peter H. "The Neuroscience of Traditional Architecture: Looking at Architectural Ornament Soothes Our Senses and Lifts Our Spirits." Traditional Building.com. Active Interest Media, 10 May 2018, https://www.traditionalbuilding.com/opinions/neuroscience-architecture. Accessed 18 Aug 2018.

Paphides, Pete. "The Music of Mathematics." Times, the (United Kingdom), n.d. https://www.thetimes.co.uk/article/delia-derbyshire-the-music-of-mathematics-mjbt5ts6g5z. Accessed 18 Aug 2018.

Parkin, Simon. "The Video Game that Maps the Galaxy." The New Yorker. Conde Nast, 9 Jul 2014, https://www.newyorker.com/tech/elements/the-video-game-that-maps-the-galaxy. Accessed 21 Aug 2018.

Pascoe, Clive B. "Motivational Pacing in Musical Design an Active Force in Music Education." Contributions to Music Education, no. 3, 1974, pp. 106–112. JSTOR, JSTOR, www.jstor.org/stable/24126929. Accessed 20 Aug 2018.

Pincus, Greg. "Greg Pincus – Writer Guy." Greg Pincus. Greg Pincus, 2013, http://www.gregpincus.com/the-14-fibs.html. Accessed 24 Aug 2018.

Posamentier, Alfred S. and Ingmar Lehmann. *The (Fabulous) Fibonacci Numbers*. Prometheus Books, NY, 2007.

Rogers, Michael R. "Chopin, Prelude in A Minor, Op. 28, No. 2." 19th-Century Music, vol. 4, no. 3, 1981, pp. 245–250. JSTOR, JSTOR, www.jstor.org/stable/746697. Accessed 20 Aug 2018.

Rory PQ, "Music and the Fibonacci Sequence w/ Rory PQ." Dubspot. DS14, Inc. – Dubspot. 1 Jun 2016. http://blog.dubspot.com/fibonacci-sequence-in-music/ Accessed 29 Oct 2018.

Salingaros, Nikos A. *Applications of the Golden Mean to Architecture in Meandering through Mathematics*, 2012.

Salingaros, Nikos A. *Why Monotonous Repetition is Unsatisfying in Meandering through Mathematics*, 2011.

Sartwell, Crispin, "Beauty", The Stanford Encyclopedia of Philosophy (Winter 2017 Edition), Edward N. Zalta (ed.), https://plato.stanford.edu/archives/win2017/entries/beauty/. Accessed 19 Aug 2018.

Shamsian, Jacob and Carl Mueller. "An Artist Uses Mathematical Formulas to Make Otherworldly Geometric Sculptures." Business Insider. Insider Inc. https://www.businessinsider.com/artist-john-edmark-uses-math-to-make-otherworldly-geometric-sculptures-2016-4?r=US&IR=T&IR=T. Accessed 19 Aug 2018.

Sinha, Sudipta. "The Fibonacci Numbers and Its Amazing Applications." International Journal of Engineering Science Invention ISSN (Online): 2319 – 6734, ISSN (Print): 2319 – 6726 www.ijesi.org ||Volume 6 Issue 9|| September 2017 || PP. 07-14, http://www.ijesi.org/papers/Vol(6)9/Version-3/B0609030714.pdf. Accessed 22 Aug 2018.

Spinak, Mike. *The Golden Section Hypothesis: A Critical Look in Naturography*. 2011.

Stakhov, Alexey. "The Mathematics of Harmony: Euclid to Contemporary Mathematics and Computer Science." World Scientific.com. World Scientific Publishing Co. Pte. Ltd., 2018, http://www.worldscibooks.com/mathematics/6635.html. Accessed 23 Aug 2018.

Stefanovic, Konstantin. "Dynamic Symmetry and Golden Section." Eng trans: Ana Smiljanic. Sanu.ac. n.d. http://www.mi.sanu.ac.rs/vismath/jadrbookhtml/part42.html. Accessed 21 Aug 2018.

Sylvestre, Loïc, and Marco Costa. "The Mathematical Architecture of Bach's 'The Art of Fugue.'" Il Saggiatore Musicale, vol. 17, no. 2, 2010, pp. 175–195. JSTOR, JSTOR, www.jstor.org/stable/43030058. Accessed 19 Aug 2018.

Van Gend, Robert. "The Fibonacci Sequence and the Golden Ratio in Music." Notes on Number Theory and Discrete Mathematics ISSN 1310–5132 Vol. 20, 2014, No. 1, 72–77, http://www.nntdm.net/papers/nntdm-20/NNTDM-20-1-72-77.pdf. Accessed 05 Sep 2018.

Weisstein, Eric W. "Golden Spiral." MathWorld. A Wolfram Web Resource. 31 Jul 2018, http://mathworld.wolfram.com/GoldenSpiral.html. Accessed 21 Aug 2018.

Wu, Feng. "Stradivarius: Music of the Golden Ratio." WufengEngineering. Wordpress, 18 Jan 2015, https://wufengengineering.wordpress.com/2015/01/18/stradivarius-music-of-the-golden-ratio/. Accessed 21 Aug 2018.

PART IV

Akhtaruzzaman, Md. and Amir A. Shafie. "Geometrical Substantiation of Phi, the Golden Ratio and the Baroque of Nature, Architecture, Design and Engineering." International Journal of Arts p-ISSN: 2168-4995 e-ISSN: 2168-5002. 2011. 1(1): 1-22. doi: 10.5923/j.arts.20110101.01. Scientific & Academic Publishing. 2012, http://article.sapub.org/10.5923.j.arts.20110101.01.html. Accessed 19 Aug 2018.

Ashrafian H, Athanasiou T. "Fibonacci Series and Coronary Anatomy." Heart Lung Circ 2011 Jul;20(7):483–4.

Bellos, Alex. "Golden ratio discovered in uterus." The Guardian. Guardian News and Media Limited, 14 Aug 2012. https://www.theguardian.com/science/alexs-adventures-in-numberland/2012/aug/14/ golden-ratio-uterus. Accessed 06 Sep 2018.

Britton, Jill. "Fibonacci Numbers in Nature." 20 Jun 2011. https://archive.li/t2RY http://britton.disted.camosun.bc.ca/fibslide/jbfibslide.htm. Accessed 29 Aug 2018.

Boeyens JCA, Thackeray JF. "Number Theory and the Unity of Science." S Afr J Sci. 2014;110(11/12), Art. #a0084, 2 pages. http://dx.doi. org/10.1590/sajs.2014/a0084. Accessed 05 Sep 2018.

Cartwright, Jon. "Fibonacci Spirals in Nature Could Be Stress-Related." SOTT.net: Signs of the Times. PhysicsWeb. 26 Apr 2007 07:13 UTC, https://sott.net/en130945, Accessed 1 Sep 2018.

Chin, Gilbert J. "Organismal Biology: Flying Along a Logarithmic Spiral." Science, 290 (5498): 1857, 8 Dec 2000, doi:10.1126/science.290.5498.1857c

Coxeter, H. S. M. *Introduction to Geometry*. Wiley: New York, 1961, p. 172.

D'Agnese, Joseph *Blockhead: The Life of Fibonacci*. Henry Holt and Co. (BYR); First edition. 30 Mar 2010, p. 40.

Davis, T. Antony. "Why Fibonacci Sequence for Palm Leaf Spirals?." (1971), http://library.isical.ac.in:8080/jspui/bitstream/10263/897/1/TFQ-9-3-1971- P237-244.pdf. Accessed 04 Sep 2018.

Devlin, K. *The Man of Numbers: Fibonacci's Arithmetic Revolution*. Walker Publishing Company, New York, 2011.

Farina, A & Fantacci, Claudio & Frasca, Marco. (2014). "Stochastic Filtering of a Random Fibonacci Sequence: Theory and Applications. Signal Processing. 104. 212–224. 10.1016/j.sigpro.2014.03.052. https://www.researchgate.net/publication/
262193339_Stochastic_filtering_of_a_random_Fibonacci_sequence_Theory_and_applications Accessed 04 Sep 2018.

Gibson CM, Gibson WJ, Murphy SA, Marble SJ, McCabe CH, Turakhia MP, et al. "Association of the Fibonacci Cascade with the Distribution of Coronary Artery Lesions Responsible for ST-Segment Elevation Myocardial Infarction." Am J Cardiol 2003;92(5):595–7.

Green, Christopher D. "All That Glitters: a Review of Psychological Research on the Aesthetics of the Golden Section." Perception. volume 24, 20 Mar 1995, 937-968 Department of Psychology, York University, North York, Ontario M3J 1P3, Canada.

Green, Christopher D. "Mathematical and Historical Background to 'All That Glitters...: A Review of Psychological Research on the Aesthetics of the Golden Section'." Department of Psychology, York University, North York, Ontario M3J 1P3 CANADA. http://www.yorku.ca/christo/papers/goldhist.htm Accessed 04 Sep 2018.

Horadam, A. F. "Eight Hundred Years Young." Vol 31, pp 123-134. The Australian Mathematics Teacher, 1975, Evansville.edu. 2018, http://faculty.evansville.edu/ck6/bstud/fibo.html. Accessed 19 July 2018.

Hutchison A. L. and R. L. Hutchison. "Fibonacci, Littler, and the Hand: a Brief Review." Hand 2010;5:364–8.

Iosa, Marco; Fusco, Augusto; and Fabio Marchetti, et al., "The Golden Ratio of Gait Harmony: Repetitive Proportions of Repetitive Gait Phases," BioMed Research International, vol. 2013, Article ID 918642, 7 pages, 2013. https://doi.org/10.1155/2013/918642. Accessed 06 Sep 2018.

Iosa, Marco; Morone, Giovanni; and Stefano Paolucci. "Golden Gait: An Optimization Theory Perspective on Human and Humanoid Walking." Frontiers in Neurorobotics. 19 Dec 2017, https://www.frontiersin.org/articles/10.3389/fnbot.2017.00069/full. Accessed 09 Sep 2018.

Jefferson, Yosh, DMD. "Facial Beauty - Establishing a Universal Standard." IJO. Vol 15, No 1, 2004, http://facialbeauty.org/article/FacialBeauty.pdf. Accessed 09 Sep 2018.

Junod, Tom. "The Falling Man." Esquire. Hearst Communications, Inc. 9 Sep 2016, https://www.esquire.com/news-politics/a48031/the-falling-man-tom-junod/. Accessed 04 Sep 2018.

Kawakami S, Tsukada S, Hayashi H, Takada Y, Koubayashi S. "Golden Proportion for Maxillofacial Surgery in Orientals." Annals of Plastic Surgery. 1989 Nov; 23(5) 417-425. PMID: 2604329. https://europepmc.org/abstract/med/2604329. Accessed 09 Sep 2018.

Knott, Ron. "A Brief Biographical Sketch of Fibonacci, His Life, Times and Mathematical Achievements." Surrey.ac.uk. University of Surrey, 25 July 2005, http://personal.ee.surrey.ac.uk/Personal/R.Knott/Fibonacci/fibBio.html. Accessed 25 July 2018.

Leppik, Elmar E. "Evolutionary Differentiation of the Flower Head of the Compositae II." Annales Botanici Fennici, vol. 7, no. 4, 1970, pp. 325–352. JSTOR, JSTOR, www.jstor.org/stable/23724678. Accessed 04 Sep 2018.

Littler JW. "On the Adaptability of Man's Hand (with Reference to the Equiangular Curve)." Hand 1973;5(3): 187–91.

"Live Algae Specimens." Niles Bio.com. Niles Biological, Inc. 7 Sep 2018, http://www.nilesbio.com/subcat103.html. Accessed 07 Sep 2018.

Livio, Mario. "The Golden Number." Natural History, vol. 112, no. 2, 03, 2003, pp. 64-69. ProQuest, https://search.proquest.com/docview/210617375?accountid=7494. Accessed 05 Sep 2018.

Majumder PP and Chakravarti A. "Variation in the Number of Ray- and Disk- Florets in Four Species of Compositae." Fibonacci Q. 14, 97–100. 1976, Google Scholar, https://www.fq.math.ca/Scanned/14-2/majumder.pdf. Accessed 04 Sep 2018.

Marco, Iosa and Morone Giovanni, Paolucci Stefano. "Golden Gait: An Optimization Theory Perspective on Human and Humanoid Walking." Frontiers in Neurorobotics Vol 11, 2017. https://www.frontiersin.org/article/10.3389/fnbot.2017.00069. DOI=10.3389/fnbot.2017.00069. Accessed 06 Sep 2018.

Masran, Mohd Rezuan. "Fibonacci Sequence and Orderliness as Observed in the Creations of Allah." Abstract. ONLINE JOURNAL OF RESEARCH IN ISLAMIC STUDIES 1.1 (2014): 104-122. https://works.bepress.com/rezuan/1/. Accessed 04 Sep 2018.

Meisner, Gary. "Golden Ratio Overview." GoldenNumber.net. PhiPoint Solutions, LLC, 12 Jul 2015, https://www.goldennumber.net/golden-ratio/. Accessed 08 Sep 2018.

Okabe, Takuya. "Physical Phenomenology of Phyllotaxis." Faculty of Engineering, Shizuoka University, 3-5-1 Johoku, Hamamatsu 432-8561, Japan, arXiv:1011.1981 [physics.bio-ph]. https://arxiv.org/pdf/1011.1981v3.pdf. Accessed 04 Sep 2018.

"The Plant List: Compositae". Royal Botanic Gardens Kew and Missouri Botanic Garden. The Plant List (2013). Version 1.1. Published on the Internet; http://www.theplantlist.org/. Accessed 04 Sep 2018.

Sacco, R.G. "Fibonacci Harmonics: A New Mathematical Model of Synchronicity." Applied Mathematics, 9, 702-718. https://doi.org/10.4236/am.2018.96048. Accessed 04 Sep 2018.

Scotta, T. C. and P. Marketos. "On the Origin of the Fibonacci Sequence." History. MCS. 23 Mar. 2014, http://www-history.mcs.st-and.ac.uk/~history/Publications/fibonacci.pdf. Accessed 19 July 2018.

Seewald, Nick. "Phyllotaxis: The Fibonacci Sequence in Nature." Golden Ratio Myth. Weebly, 2010, https://goldenratiomyth.weebly.com/phyllotaxis-the-fibonacci-sequence-in-nature.html. Accessed 04 Sep 2018.

Shabalkin, I.P., Grigor'eva, E. Yu., Gudkova, M.V., and Shabalkin, P.I. "Fibonacci Sequence and Supramolecular Structure of DNA." 20 May 2014. Translated from Kletochnye Tekhnologii v Biologii i Meditsine, No. 1, pp. 60-64, January, 2016, https://search.proquest.com/docview/1860892882/C12EB9450A12478APQ/1?accountid=7494. Accessed 05 Sep 2018.

Sinha, Sudipta. "The Fibonacci Numbers and Its Amazing Applications." International Journal of Engineering Science Invention ISSN (Online): 2319 – 6734, ISSN (Print): 2319 – 6726 www.ijesi.org ||Volume 6 Issue 9|| September 2017 || PP. 07-14, http://www.ijesi.org/papers/Vol(6)9/Version-3/B0609030714.pdf. Accessed 06 Sep 2018.

Spooner, David. *The Poem and the Insect: Aspects of Twentieth Century Hispanic Culture*. UPA 19 Mar 2002. pp. 37-38.

Swinton J, Ochu E. "Novel Fibonacci and Non-Fibonacci Structure in the Sunflower: Results of a Citizen Science Experiment." MSI Turing's Sunflower Consortium. R Soc Open Sci. 2016 May; 3(5):160091. Epub 2016 May 18. National Center for Biotechnology Information, U.S. National Library of Medicine, 2018, https://www.ncbi.nlm.nih.gov/pubmed/27293788. Accessed 04 Sep 2018.

Tracy, Suzanne, and HPC Source. "Celebrating Fibonacci Day 2015 on Monday, November 23." Scientific Computing, 2015. ProQuest, https://search.proquest.com/docview/1739102165?accountid=7494. Accessed 05 Sep 2018.

Trinajstić, Nenad. "The Magic of the Number Five." Croatica Chemica Acta 66.1 (1993): 227-254.

Tung, K.K. *Topics in Mathematical Modeling*. Princeton: Princeton University Press. 2007. Print.

Verguts, J.; Ameye, L.; Bourne, T. and Timmerman, D. (2013), "Normative Data for Uterine Size According to Age and Gravidity and Possible Role of the Classical Golden Ratio." Ultrasound Obstet Gynecol, 42: 713-717. doi:10.1002/uog.12538

Wright, Michael Lowe. "The Golden Mean." Vashti.net. Michael's Crazy Enterprises, Inc., 2017, http://www.vashti.net/mceinc. Accessed 04 Sep 2018.

Yetkin, G., Sivri, N., Yalta, K., and Yetkin, E. "Golden Ratio is Beating in Our Heart." Int J Cardiol. 2013 Oct 12; 168(5):4926-7. http://dx.doi.org/10.1016/j.ijcard.2013.07.090. Accessed 06 Sep 2018.

a note about the author

Shelley Allen is a writer, editor, poet, and mother of four. She holds a Bachelor of Arts degree in English and a Master's Degree in Secondary Education from Southern Methodist University. A former teacher and bookstore owner, Allen is a bibliophile with broad literary interests. She works professionally as a freelance writer from her home in Dallas, Texas.

This book was designed, edited, and set into type
by Tarek I. Saab.

The text face is Iowan Old Style,
designed by John Downer
and issued in digital form by Bitstream in 1991.

The paper is acid-free and is of archival quality.